CLASSIQUEST SCIENCE LOGIC STAGE BIOLOGY

Angela DuBois

Classical Education Resources, LLC

Sherrills Ford, NC 28673

www.classicaleducationresources.com

© 2010 Classical Education Resources, LLC.   All rights reserved.

Copyright © 2010 Classical Education Resources, LLC. All Rights Reserved

ISBN-13: 978-0-9829573-1-8

ISBN-10: 0-9829573-1-9

Version 1.0

First Printing, October 2010

No part of this publication may be reproduced, stored in a retrieval system or transmitted in any form, nor by any means including but not limited to electronic, mechanical, copy, print, recording, or otherwise, without the express written permission of the author. The purchaser of this study guide may copy the designated reproducible forms and diagrams for use with his or her own family. Reproduction for group, classroom, school or co-operative use is prohibited. Discounted purchasing is available for non-family groups. Contact Classical Education Resources for further information.

The publisher and author have made every reasonable effort to insure that the experiments and activities in the book are safe when conducted as instructed but assume no responsibility for any damage caused or sustained while performing the experiments or activities in this book. Instructors, parents, and/or guardians should supervise young readers who undertake the experiments and activities in this book.

Disclaimer of Warranty and Limit of Liability: While the publisher and author have used their best efforts in preparing this book, they make no representations or warranties with respect to the accuracy or completeness of the contents of this book and specifically disclaim any implied warranties of merchantability or fitness for a particular purpose. No warranty may be created or extended by sales representatives or written sales materials. The information contained herein may not be suitable for your situation. You should consult with a professional where appropriate. Neither the publisher nor author shall be liable for any loss of profit or any other damages, including but not limited to special, incidental, consequential, or other damages.

Published by Classical Education Resources, Sherrills Ford, NC 28673

www.classicaleducationresources.com

For Chase

Quaere verum - sapere aude.

Seek truth – dare to be wise.

Table of Contents

Preparation 11

    Using ClassiQuest Science 11

    The Science Notebook 15

    Resources and Materials 17

    Safety Guidelines 31

Fundamentals 33

    The Scientific Method – 'Doing' Science 33

    Using the Microscope 37

Life in General 43

    Lesson 01: The Scientific Method 43

    Lesson 02: Classifying Life 45

    Lesson 03: Life and Its Cycles 47

    Lesson 04: The Microscopic World 49

    Lesson 05: Cells: The Building Blocks of Life 51

    Lesson 06: Reproduction 53

    Lesson 07: Cell Division 55

    Lesson 08: Genetics 57

© 2010 Classical Education Resources, LLC.   All rights reserved.

Lesson 09: Simple Organisms 59

Lesson 10: Neither Plant nor Animal 61

Botany – The Study of Plants 63

Lesson 11: Structure of Plants 63

Lesson 12: Plant Leaves and Food Production 65

Lesson 13: Plant Sensitivity 67

Lesson 14: Flowers 69

Lesson 15: Fruits and Seeds 71

Lesson 16: Vegetative Reproduction 73

Zoology - The Study of Animals 75

Lesson 17: Mollusks, Cnidaria & Echinoderms 75

Lesson 18: Worms 77

Lesson 19: Arthropods: Arachnids and Crustaceans 79

Lesson 20: Arthropods: Insects 81

Lesson 21: Fish 83

Lesson 22: Amphibians 85

Lesson 23: Reptiles 87

Lesson 24: Birds 89

Lesson 25: Mammals 91

Human Anatomy 93

Lesson 26: Respiratory System 93

Lesson 27: Circulatory System 95

Lesson 28: Digestive System 97

Lesson 29: Muscular System 99

Lesson 30: Skeletal System 101

Lesson 31: Nervous System 103

Lesson 32: Skin and Touch 105

Lesson 33: Taste and Smell 107

Lesson 34: Ears and Hearing 109

Lesson 35: Eyes and Sight 111

Ecology – The Big Picture *113*

Lesson 36: The Real World Wide Web 113

ClassiQuest Labs *115*

Lab Diagrams *171*

Memorization Lists *183*

List A: The Seven Major Groups of Classical Taxonomy 183

List B: The Five Kingdoms of Classical Taxonomy 184

List C: Parts of a Flower 185

List D: Animal Kingdom Phyla 186

List E: Major Systems of the Human Body 187

List F: Major Bones of the Human Body 188

List G: Major Biomes of the Earth 189

Forms *193*

Lab Report: Objective & Materials 193

Lab Report: Procedure 194

Lab Report: Hypothesis 195

Lab Report: Results, Narrative 196

Lab Report: Results, Data 197

Lab Report: Results, Microscopy 198

Lab Report: Results, Observation 199

Lab Report: Conclusion 200

Important Dates 201

Timeline 205

Terminology from Latin & Greek Roots 207

Answer Key 213

Thinking Critically About Science 229

INTRODUCTION

USING CLASSIQUEST SCIENCE

ClassiQuest Science study guides provide a solid science foundation for those using the classical method of education.

ClassiQuest Science is sequenced according to the trivium of the classical model.[1] There are different approaches to scheduling the trivium. The structure used here divides the educational odyssey into three cycles of four years each, known as the *grammar stage*, the *logic stage* (or *dialectic stage*) and the *rhetoric stage*.

The grammar stage encompasses years one through four and concerns itself with the basics of the subjects studied. Years five through eight comprise the logic stage in which the student develops the ability to apply critical analysis to the material encountered. The rhetoric stage consists of years nine through twelve and requires the student to effectively synthesize and communicate.

Some subjects are subdivided using history as the core organizing principle. In the first year of each trivium stage, focus is given to the Ancients. The second year is spent on the Medieval period, the third on the Early Modern era, and the last on Modern times. This cycle is repeated a total of three times during the twelve preparatory years. The correspondence to modern western grade levels is illustrated below:

	Grammar	Logic	Rhetoric
Ancient	1st grade	5th grade	9th grade
Medieval – Early Renaissance	2nd grade	6th grade	10th grade
Late Renaissance – Early Modern	3rd grade	7th grade	11th grade
Modern	4th grade	8th grade	12th grade

---

[1] For a thorough treatment of the subject of classical education, see The Well-Trained Mind: A Guide to Classical Education at Home. Bauer, Susan Wise and Jessie Wise. 3rd ed. New York: W. W. Norton & Company, 2009.

In the grammar stage, the objective of science study is to gain exposure through exploration. In the logic stage students develop critical thinking through the use of the scientific method, careful observations, and organizational activities. The rhetoric stage develops the rigorous skills required of empirical science while enlightening the student to the development of scientific thought through the reading of original sources.

Just as a different period of history is studied each year of the stage, a different scientific discipline gets focus. The branch of science chosen is the one that generally dominated intellectual inquiry during the period. Thus, in the first year of each stage, when Ancient history is being examined, the student studies the life sciences. The Medieval period is spent on the geosciences and astronomy, the Renaissance to Early Modern period brings chemistry to the forefront and the Modern period attends to physics. The branches are studied three times during the twelve years, each time at a more advanced level, corresponding to the student's growing abilities.

The chart below summarizes the years of science study:

Period	Science	Years
Ancient	Biology	1st, 5th, 9th grades
Medieval – Early Renaissance	Astronomy & Geosciences	2nd, 6th, 10th grades
Late Renaissance – Early Modern	Chemistry	3rd, 7th, 11th grades
Modern	Physics	4th, 8th, 12th grades

LESSONS

The lessons in this study guide are divided into six sections as follows:

Reading & Research –Readings are assigned from high quality reference works. As often as possible, two or more sources are consulted.

Definitions – Students provide the precise definitions required of science while simultaneously being exposed to the Latin and Greek roots of scientific terminology.

Lab Activities – Students use critical thinking skills while developing a rigorous approach to problem-solving. Lab Reports encourage orderly record keeping.

Report – Students are required to write about what they have learned. This serves both to cement the information in the student's mind and to build the skill of articulate communication.

Science History – Science is made globally and historically relevant through the use of date references and a timeline.

Memory Work – Students practice recall and begin to learn speaking skills by performing oral recitation.

Note: For the sake of clarity and brevity the book titles are abbreviated in the lessons as indicated below:

KSE   The Kingfisher Science Encyclopedia

USE   The Usborne Internet-Linked Science Encyclopedia

DSE   Encyclopedia of Science, DK Publishing

DB   Usborne Illustrated Dictionary of Biology

BEK   Biology for Every Kid

See the Resources and Materials list for more information about these books.

LABS

The forms needed for the course are provided in the rear of the study guide. You are permitted to make as many copies of these forms as you need for use within your own family. Several Lab Diagrams are provided following the labs section and these should also be copied.

The labs used with the lessons are found either in the Labs section of this study guide or in the book Biology for Every Kid by Janice VanCleave.

In the early labs, most of the lab reporting has already been done for the student, including some custom data collection forms. If you wish, you may copy the ClassiQuest lab pages and file the copies in the Science Notebook with the student created forms so that the lab reports are complete. For the labs coming from the VanCleave book, the student will write the entire report, copying the Objective, Materials and Procedures sections onto Lab Report forms.

Occasionally, the student will need to construct a graph, table or some other structure for collecting data. If possible, let the student think through the type of format best suited for the task. Encourage them to try some sketches on scrap paper to better visualize the options. If the student becomes frustrated or overwhelmed, by all means offer help. At this stage, they may still need assistance deciding which format is best for collecting various kinds of information.

At the end of the study guide is a section called "Thinking Critically About Science." It is, by necessity, somewhat philosophical, but is written in a conversational style appropriate for logic stage students. Instructors are encouraged to read it and then assign it to students along with the first lesson.

THE SCIENCE NOTEBOOK

A notebook is used to organize the student's work. A standard three-ring binder is ideal for this purpose. It should be divided into 7 sections as follows:

- Research Reports
- Definitions
- Lab Activities
- Diagrams
- Dates
- Memory Work
- Extra Activities

Research Reports – The student will write a brief report on the material covered in each lesson and file it in this section. These reports may be written on notebook paper or typed on a computer and punched to fit the binder.

Definitions – Each lesson asks the student to define a few terms. These definitions should be written on notebook paper and filed in this section. Most of the definitions can be found in the reading assignments, but a dictionary will be needed for some of them. Below the definition, the student should also note any Latin or Greek word parts. For example:

> biology – the study of life
>
> - bio – life (Greek)
> - logy – study of (Greek)

A list of some common Latin and Greek word parts is included in this guide for reference.

Lab Activities – Each lesson has a coordinated lab activity. Forms are included to copy and use for the Lab Reports, which will be filed in this section of the notebook.

Diagrams - Some of the lessons include diagrams to label. The student may also wish to make sketches of models or observations. These types of documents will be filed in the Diagrams section of the notebook.

Dates – Significant dates are to be recorded in this section of the notebook. Most lessons suggest important dates which can be found in the assigned readings or the various appendices of the reference books used. Reproducible forms are included for recording these dates. The forms have headings for each level of the logic stage – Ancient, Medieval to Early Renaissance, Late Renaissance to Early Modern, and Modern. You should begin with at least one copy of each of the forms placed in the Dates section of the notebook. As dates of scientific interest are encountered, they should be recorded on the appropriate page

If a timeline is being constructed for history studies, the dates should also be recorded there. Otherwise, a science timeline can be created to give perspective to the important discoveries and developments encountered. A timeline form is included in this guide. You will need to make several copies of this form and glue or tape the pages together end to end as shown below (3 copies are shown).

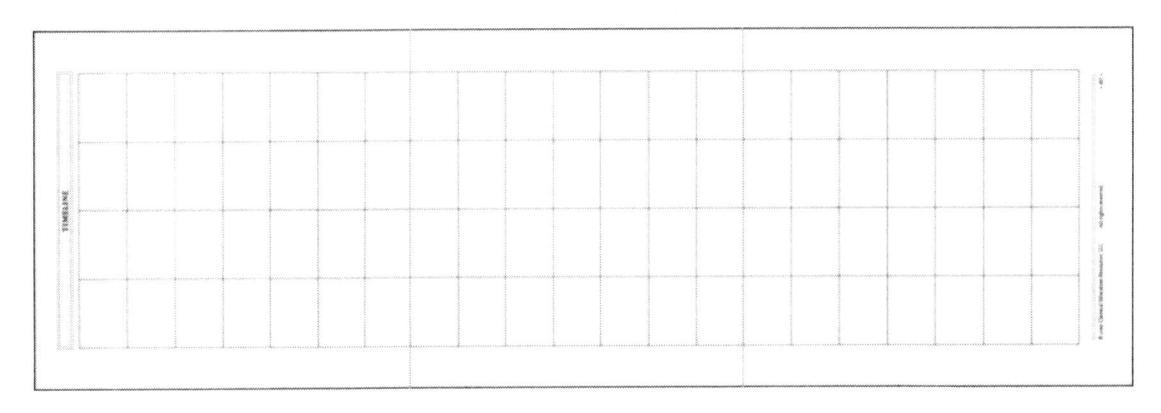

When turned on its side, this form provides 7 sections per page. It is recommended that you use one section to represent a 100 year increment. If dating is begun at 5000 BCE 11 copies will be needed. The form also has four rows that may be used in several ways. They could divide the timeline into fields of study (biology, astronomy and geosciences, chemistry, physics), into types of entries (thinkers, discoveries, written works, inventions), or the rows can simply be ignored.

Memory Work – The memorization lists should be copied and filed in this section where they will be available for frequent review.

PREPARATION

RESOURCES AND MATERIALS

BOOK LIST *(used in all four years of Logic Stage science)*

The Kingfisher Science Encyclopedia (KSE)

The Usborne Internet-Linked Science Encyclopedia (USE)

Encyclopedia of Science, DK Publishing (DSE)

*The Usborne Illustrated Dictionary of Science (DB) (all 4 years of the Logic Stage)

    or The Usborne Illustrated Dictionary of Biology (this year only)

LAB MATERIALS

Biology for Every Kid, VanCleave, Janice (BEK)

Home Science Tools Home Microscope

    or Thames and Kosmos TK2 Microscope and Biology Kit

    or comparable microscope   (will also be used in high school courses)

Botanical Discoveries Kit, DuneCraft (BDK)

Owl Pellet Dissection Kit, Home Science Tools or any brand (OPK)

Blood Typing Kit, Home Science Tools or any brand (BTK)

Microscope Slide-making Kit or equivalent (MSK), Home Science Tools or any brand

    Eosin Y stain

    Methylene blue stain

    Pipette

    Plain slides, 12

    Concavity slides, 2

    Coverslips

    Storage box for slides

---

*This work contains 3 volumes: Usborne Illustrated Dictionary of Biology, Usborne Illustrated Dictionary of Chemistry, and Usborne Illustrated Dictionary of Physics.

PREPARATION

Prepared microscope slides (SLD) available from Home Science Tools and other sources

    hydra budding

    frog sperm

    frog ovary

    allium (onion) root tip showing mitosis

    lilium (lily anther) showing meiosis

    flower buds, 2 types - monocot and dicot

    insect legs, 4 types

    fish scales, 3 types – cycloid, ctenoid, placoid (only cycloid is required)

    muscle, 3 types – striated, smooth, cardiac

Uncover a Tarantula book/model, Silver Dolphin Books

Uncover a Frog book/model, Silver Dolphin Books

Uncover a Cobra book/model, Silver Dolphin Books

It is recommended that you gather all non-living/non-perishable supplies ahead of time. A large (15 quart or greater) plastic tub with lid is well suited for this purpose.

Some materials will need to be obtained from a science supply company such as Home Science Tools or Nasco. Many others can be found by 'shopping' around the house. For the rest, discount superstores can be a convenient choice since they carry a wide variety of materials at reasonable prices.

PREPARATION

MATERIALS LIST (*per lab*)*

L= Living/ Perishable

Lab	Item *(kit)*	Alternate	Qty	L
01	mustard seeds (BDK)		6-8	
01	sensitive plant seeds (BDK)		6-8	
01	cacti seeds (BDK)		6-8	
01	germination mixture (BDK)		1 ½ c	
01	germination cups (BDK)		3	
01	germination cup lids (BDK)		3	
01	plant stakes (BDK)		3	
01	fine tip marker			
02	colored pencils			
02	scissors			
03	dried yeast		3 t	
03	sugar		2 t	
03	clear disposable cups, 16oz		3	
03	craft ticks/tongue depressors		2	
03	measuring cup		¼	
03	measuring spoon, teaspoon		1 t	
03	masking tape			
03	fine tip marker			
03	ruler, metric & English			
03	timer or clock			
04	microscope			

*Abbreviations:
BDK-Botanical Discoveries Kit, DuneCraft
OPK-Owl Pellet Dissection Kit, Home Science Tools or any brand
BTK-Blood Typing Kit, Home Science Tools or any brand
MSK-Microscope slide-making kit or equivalent available from Home Science Tools

© 2010 Classical Education Resources, LLC.   All rights reserved.

PREPARATION

04	blank slide (MSK)			
04	coverslip (MSK)			
04	cork			
04	single edge razor blade			
04	tweezers (OPK)			
04	pipette (MSK)			
04	colored pencils			
05	microscope			
05	blank slides (MSK)		2	
05	coverslips (MSK)		2	
05	eosin Y stain (MSK)			
05	methylene blue stain (MSK)			
05	tweezers (OPK)			
05	pipette (MSK)			
05	clear disposable cup, 16oz			
05	small onion			L
05	colored pencils			
05	tongue depressor or craft stick			
06	microscope			
06	prepared slide, hydra budding			
06	prepared slide, frog sperm			
06	prepared slide, frog ovary			
06	colored pencils			
07	microscope			
07	prepared slide, allium (onion) root tip showing mitosis			
07	prepared slide, lilium (lily anther) showing meiosis			
07	colored pencils			
09	table salt		1 t	

© 2010 Classical Education Resources, LLC.   All rights reserved.

~ 20 ~

PREPARATION

09	white vinegar		1 t	
09	small clear glasses	clear disposable cup, 16oz	3	
09	chicken bouillon cube			
09	measuring cup			
09	measuring spoon, teaspoon			
09	masking tape			
09	marking pen	fine tip marker		
10	whole mushroom from the store (not wild! – some are poisonous)			L
10	dark construction paper			
10	disposable bowl			
10	art fixative or hairspray			
10	knife			
11	measuring cup			
11	drinking glasses	clear disposable cups, 16oz	2	
11	white carnation with a long stem			L
11	food coloring, red			
11	food coloring, blue			
12	house plant	mustard plant (BDK)		L
12	black construction paper			
12	scissors			
12	tape (cellophane)	masking tape		
13	shoebox with lid	(any similar box with lid)		
13	paper cup	germination cup (BDK)		
13	pinto beans	mustard plant seeds (BDK)	3	L
13	potting soil	germination mixture (BDK)		
13	cardboard		2 pcs	
13	scissors			
13	tape	masking tape		
14	microscope			

© 2010 Classical Education Resources, LLC.   All rights reserved.

PREPARATION

14	prepared slide, 2 flower types			
14	colored pencils			
15	pinto beans		10-12	
15	jar	plastic gallon jar with lid		
15	paper towels			
16	potatoes		4	L
16	potting soil	germination mixture (BDK)		
16	quart jar (liter jar)	plastic gallon jar with lid		
17	suction cup(s)			
17	smooth rock(s)			
18	earthworms			L
18	paper towel			
18	cotton balls		1	
18	nail polish remover		¼ oz	
19	Uncover a Tarantula book/model			
20	microscope			
20	prepared slide, 4 insect legs			
20	colored pencils			
21	microscope			
21	prepared slide, fish scales			
21	colored pencils			
22	Uncover a Frog book/model			
23	Uncover a Cobra book/model			
24	owl pellet (OPK)			
24	tweezers (OPK)			
24	probe (OPK)			
24	guide and worksheet (OPK)			
24	bone sorting chart (OPK)			
24	skull sorting key (OPK)			

© 2010 Classical Education Resources, LLC. All rights reserved.

PREPARATION

24	disposable gloves			
24	paper plates, 3	disposable dissecting pans	2	
25	microscope			
05	blank slides (MSK)		3	
05	coverslips (MSK)		3	
05	tweezers (OPK)			
05	pipette (MSK)			
25	human hair			
25	dog hair			
25	cat hair			
25	colored pencils			
26	plastic dishpan	clear plastic storage tub w/lid, approx 15qt		
26	aquarium tubing	vinyl tubing	2 ft	
26	plastic milk jug with cap, 1 gallon size	plastic gallon jar with lid		
26	masking tape			
26	pen	fine tip marker		
27	blood typing card (BTK)			
27	pipette (BTK)			
27	alcohol prep pad (BTK)			
27	sterile lancet (BTK)			
27	mixing sticks, 4 (BTK)		4	
27	disposable gloves			
28	sugar		1 Tbsp	
28	cornstarch		1 Tbsp	
28	tincture of iodine, colored			
28	funnel			
28	round coffee filters	cone type coffee filters	5	
28	glass jar	plastic gallon jar with lid		

© 2010 Classical Education Resources, LLC.   All rights reserved.

PREPARATION

28	small drinking glass	clear disposable cup, 16oz		
28	measuring cup			
28	measuring spoon, tablespoon			
28	eye dropper	pipette (BTK)		
29	microscope			
29	prepared slide, 3 muscle types			
29	colored pencils			
30	uncooked, thin chicken bone, wing or wishbone			L
30	jar with lid	plastic gallon jar with lid		
30	white vinegar		½ - 1 c	
31	pencils, sharpened		2	
31	masking tape			
32	outdoor thermometer			
32	cotton ball		1	
32	rubbing alcohol		½ oz	
32	timer or clock			
33	toothpicks		2-4	
33	blindfold			
33	spring-type clothespin	two fingers		
33	apple			L
33	onion			L
34	metal spoon			
34	kite string	twine or heavy string	2 ft	
34	ruler, metric & English			
35	magnifying lens			
35	sheet of typing paper	white paper, no lines		
35	ruler, metric & English			
36	clear glass bowl, 1 qt	clear plastic storage tub w/lid, approx 15qt		

© 2010 Classical Education Resources, LLC.   All rights reserved.

PREPARATION

36	measuring cup			
36	liquid oil	cooking oil	1 t	
36	powdered laundry detergent		2 t	
36	measuring spoon, teaspoon			

PREPARATION

CATEGORIZED MATERIALS LISTS

Science Equipment and Materials
Available from Home Science Tools and other science supply companies

Lab(s)	Item	HST #
4,5,6,7,14,20,21,25,29	microscope	MI-4100STD
4,5	Microscope Slide-Making Kit	MS-KIT04
1,12,13,16,	Botanical Discoveries Kit	KT-DCBOTAN
4,5,24	Owl Pellet Kit	PM-OWLKIT
27	Blood Typing Kit	BE-BLDTEST
6	prepared slide, hydra budding	MS-HYDRBUD
6	prepared slide, frog sperm	MS-FRSPERM
6	prepared slide, frog ovary	MS-FROVARY
7	prepared slide, allium root tip	MS-ALLROOT
7	prepared slide, lilium	MS-LILANTH
14	prepared slide, 2 flower types	MS-FLOWER
20	prepared slide, 4 insect legs	MS-INSLEGS
21	prepared slide, fish scales	MS-FISCALE
29	prepared slide, 3 muscle types	MS-MUSC3

Common Items Available from Home Science Tools
Available from around the house or Home Science Tools/science supply company or your local superstore/retail store

Lab(s)	Item	Qty	HST #
2,4,5,6,7,14,20,21,25,29	colored pencils	6-12	BE-CPENCIL
3,5	tongue depressors or craft sticks	4	GS-TONGDEP
3	dried yeast	2	LM-YEAST
3,32	timer		ME-STOPWAT
3,34,35	ruler, metric & English		ME-RULER
4	cork	1	CE-CORK09

© 2010 Classical Education Resources, LLC.   All rights reserved.

PREPARATION

4	single edge razor blade	1	CE-RAZOR
10,12,35	dark & white paper	2ea	GS-PAPERCS
11	food coloring, red & blue		CH-FOODCOL
15,16,26,28,30+storage	plastic gallon jar with lid		BE-JAR1GAL
24	dissection tray	2	DE-TRAYS
24,27	disposable gloves	2pr	GS-GLOVNIT
26	aquarium tubing/vinyl tubing	2 ft	CE-TUBEPLS
28	funnel		CE-FUNN8OZ
28	tincture of iodine, colored		CH-IODINE
32	outdoor thermometer		ME-THERMME
34	kite string/cord	2 ft	MC-STRING
35	magnifying lens		OP-MAGNIFY

Common Items

Available from around the house or your local superstore/retail store

Lab(s)	Item	Qty
1,3,9,26	fine tip marker	
2,12,13	scissors	
3,5,9,11,28	clear disposable cups, 16 oz	10
3,9,36	measuring spoon, tsp	
3,9,11,28,36	measuring cup, 2 c	
3,9,12,13,26,31	masking tape	
10	disposable bowls	1
3,28	sugar	5 t
9	table salt	1 t
9	chicken bouillon cube	
9,30	white vinegar	1 - 2 c
10	knife	
10	art fixative or hairspray	

© 2010 Classical Education Resources, LLC.   All rights reserved.

PREPARATION

13	shoebox with lid or craft box	
13	cardboard	2 pcs
15	pinto beans	15
15,18	paper towels	
17	suction cup(s)	
17	smooth rock(s)	
18	nail polish remover	¼ oz
18,32	cotton balls	2
26,36 + storage	clear plastic storage tub w/lid, approx 15qt	
28	measuring spoon, Tbsp	
28	cornstarch	1 T
28	round coffee filters, full size	5
31	pencils, sharpened	2
32	rubbing alcohol	½ oz
33	blindfold	
33	toothpicks	2-4
33	spring-type clothespin or fingers	
34	metal spoon	
36	cooking oil	1 t
36	powdered laundry detergent	2 t

Living/Perishable Materials

These items require a little advance planning, but not too far in advance!

Lab	Item	Qty
05	small onion	
10	whole mushroom from the store (not wild! – some are poisonous)	
11	white carnation with a long stem	
12	house plant - grow mustard plant from Botanical Discoveries Kit (6-8 seeds)	
13	pinto beans or mustard seeds from Botanical Discoveries Kit (6-8 seeds)	

PREPARATION

16	potatoes	4
18	earthworms	4-6
30	uncooked, thin chicken bone (wing or wishbone)	
33	apple	
33	onion	

Book/Models
Available at Rainbow Resource Center, Amazon, and many booksellers

Lab(s)	Item	HST #
19	Uncover a Tarantula book/model	
22	Uncover a Frog book/model	BK-UNCFROG
23	Uncover a Cobra book/model	

Optional convenience items
To gather and store cooking oil, cornstarch, pinto beans, etc

Lab(s)	Item	Qty	HST #
27,28	pipettes, 10-pack	1	CE-PIPET
9,18,30,32,	bottle for liquids, 250ml	4	CE-BTP250N
3,9,28,36	zip bags for powders, beans, etc, 5 x 7	6-10	PK-ZIP5X7
3,9,18,28,30,32,36	labels for bottles and zip bags	8	

© 2010 Classical Education Resources, LLC.   All rights reserved.

SAFETY GUIDELINES

Instructor supervision is necessary and required for all experiments!

Some common-sense safety guidelines:

Always follow manufacturer's directions precisely.

Wear goggles, gloves and a splash apron or old clothes.

Stains used for dyeing specimens can present health hazards. Follow manufacturer's directions for use, clean up and disposal.

Sharp items like razor blades and lancets present an injury risk. Be sure to keep in a protective covering when not in use.

Glass slides and coverslips can be sharp, and can break easily. Take reasonable precautions with glass items.

Some plant matter can be toxic, or present an allergic risk to some people. Be sure to wear gloves and wash hands thoroughly.

Food items used in experimentation should never be consumed and must be kept separate from other food that will be eaten.

Always clean up when finished with lab work. Put all materials out of the reach of young children and household pets.

THE SCIENTIFIC METHOD – 'DOING' SCIENCE

The scientific method is the logical approach scientists use to answer questions. It has evolved over thousands of years. Many thinkers have written on the topic of applying critical thinking, including Sir Francis Bacon (1561-1626) in <u>Novum Organum</u> (1620) and Rene Descartes (1596-1650) in his <u>Discourse on the Method</u> (1637).[2] By following a consistent set of steps, scientists can duplicate each other's research, verify the accuracy of conclusions, and advance fields of scientific knowledge.

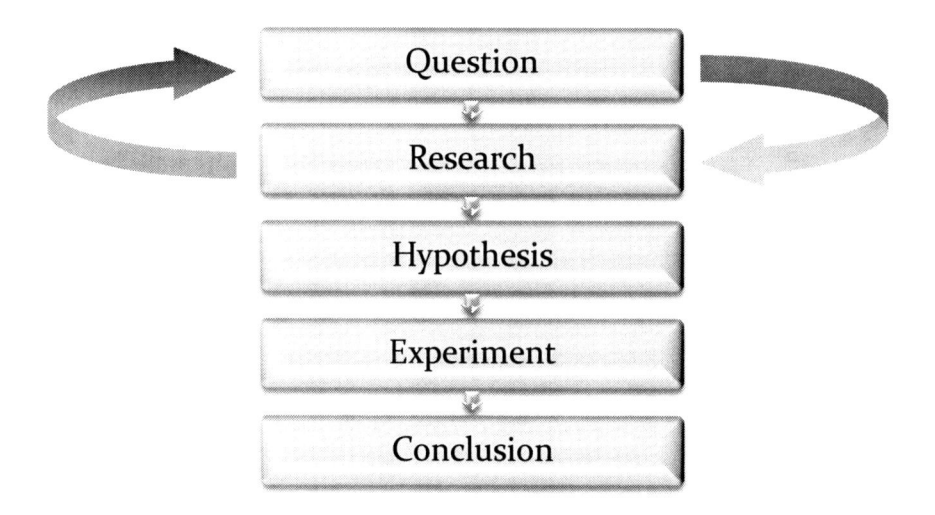

```
      ┌──────────────┐
  →   │   Question   │   ↘
      └──────┬───────┘
             ↓
      ┌──────────────┐
      │   Research   │   ←
      └──────┬───────┘
             ↓
      ┌──────────────┐
      │  Hypothesis  │
      └──────┬───────┘
             ↓
      ┌──────────────┐
      │  Experiment  │
      └──────┬───────┘
             ↓
      ┌──────────────┐
      │  Conclusion  │
      └──────────────┘
```

A simplified form of the scientific method utilizes 5 core steps:

1. **Question**: The path of scientific inquiry usually begins with a question. The question may be stimulated by an observation in the laboratory or in nature, or it may occur as the result of some information obtained through reading.

2. **Research**: Once a question has been formulated, the researcher will usually do further reading to learn more about the topic.

---

[2] The complete title is <u>Discourse on the Method of Rightly Conducting One's Reason and of Seeking Truth</u>

3. **Hypothesis**: When enough background information has been gathered, a *hypothesis* can be made. The hypothesis is an educated guess based on what's been learned so far.

4. **Experiment**: After forming the hypothesis, the researcher will perform an experiment to test the prediction. The experiment must follow certain rules in order to adhere to the scientific method.

5. **Conclusion**: Finally, the researcher will engage in a thorough analysis of the data gathered during the experiment. This analysis will lead to a conclusion. Either the hypothesis is supported by the evidence, or it is not. Either result is a valid end-product of experimentation because it provides valuable information. This information can be used to design further research and experimentation.

THE EXPERIMENT

A well-designed experiment contains two types of *variables*. Variables are conditions that change. These variables should be measurable either as an amount or as a yes/no condition.

The condition being tested is called the *independent variable*. It is the variable that is changed by the experimenter. An experiment should contain **only one** independent variable. This helps the experimenter be sure that any results obtained were caused by the condition being tested.

The condition that changes as a result of the independent variable is called the *dependent variable*. It is the factor that is measured by the experimenter. A good dependent variable is specific and can be measured.

A good experiment also has a *control*. The control is another setup that is exactly the same as the experimental one, except that it does not have the independent variable. It is used as a basis for comparison. For instance, if the experiment is testing the effect of the absence of light on mimosa plants, the control would be given normal light amounts. This would allow the experimenter to see what happens in ordinary circumstances.

FUNDAMENTALS

The hypothesis can usually be expressed as a cause and effect statement that relates the independent and dependent variables.

> *If* <independent variable> *then* <dependent variable>.

The hypothesis for the mimosa plant experiment above might be expressed like this:

> *If* the plant does not get light *then* it will show less growth.

The independent variable is the amount of light, in this case a yes/no condition (no light) and the dependent variable is the measurable amount of growth.

Points to remember:

1. It is important that you have only one independent variable. Introducing more than one independent variable prevents you from drawing a clear cause and effect relationship between your independent variable - the test - and your dependent variable – your result.
1. It is important that your variables are measurable. Some variables are easily quantifiable, like height and time. Others are *qualitative*, like the amount of green in a leaf. You will need a method of assessing and reporting those conditions consistently, such as a color scale for comparison.
2. The procedure should be well documented. All materials used and steps taken should be written down so that the results can be duplicated by others.
3. All data should be accurately documented. If there are errors in data collection or unexpected mishaps, they should be honestly reported.
4. For long-term projects such as science fairs, it is valuable to repeat the experiment several times. You want to make sure that any result you receive is not just an accident that occurs randomly, but is related to the independent variable.
5. Sometimes the hypothesis is not supported. It is important not to change your hypothesis in response to unexpected results. An unsupported hypothesis is one of the possible results of an experiment. You may choose to repeat the original experiment, redesign the experiment, or test a new hypothesis. Either way, if the scientific method has been followed, you've obtained valuable information and you should report it.

DEFINE 'PROOF'

Just how sure is sure? Suppose that someone hands you a sealed metal box and tells you it's full of green marbles. The box cannot be opened but has a trap door that will release one marble at a time. How would you test the assertion that the box is full of green marbles?

If you eject a marble and it's green, have you proven that it was a box full of green marbles? Of course not; you have simply shown that it did contain at least *one* green marble. What happens if you release another marble and it's red? You have disproved the assertion that it was a box full of green marbles.

What if you release all the marbles and they are all green. You shake the box and do not hear any more marbles rolling around. Have you proven that it was, in fact, a box full of green marbles? It might seem that way. But what if someone has firmly glued a red marble to the inside of the box so that it cannot roll around? It might have seemed to you that all the marbles were green, but there was a factor you did not know about – the undiscovered red marble.

Hypotheses work in much the same way. Since we cannot possibly know every factor that might affect the outcome of the experiment, we cannot actually *prove* a hypothesis. We can only accept it as supported or reject it as unsupported.

How, then, do scientists actually know anything? They keep trying new ways to disprove hypotheses. If enough scientists perform the same experiment and cannot disprove the hypothesis, it may become a *scientific theory*. Even then, it may one day be rejected when new information comes to light.

FUNDAMENTALS

USING THE MICROSCOPE

Before getting started, it is important to read the documentation that came with your microscope. Carefully follow all safety and care guidelines. Once you've familiarized yourself with the microscope you will be using, you can proceed to your first observation.

Most student microscopes have the same basic parts. Below is a diagram of a standard compound microscope.

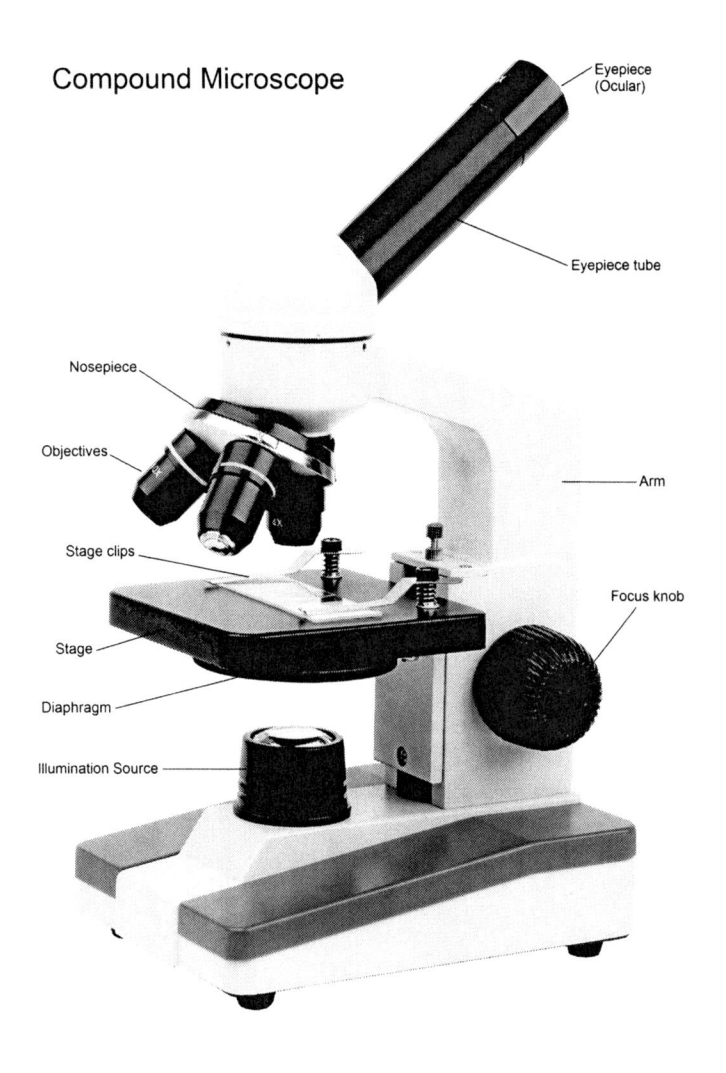

Compound Microscope

Labeled parts: Eyepiece (Ocular), Eyepiece tube, Nosepiece, Objectives, Arm, Stage clips, Focus knob, Stage, Diaphragm, Illumination Source

The *stage* is the platform on which you place the slide. It has two clips to hold the slide in place. A clip can be raised by pressing down on the back of it allowing you to easily slip the slide underneath. Some microscopes have an optional *mechanical stage*. The mechanical stage allows you to move the slide around easily.

The microscope contains two kinds of lenses. The *ocular lens* is the one closest to the eye, and it usually provides 10x magnification. The *objective lenses* are the ones closer to the object being viewed. Most student microscopes have three objective lenses with magnifications of 4x, 10x and 40x. The total magnification is found by multiplying the ocular lens magnification by the objective lens magnification. Thus, the total magnifications provided are 40x, 100x and 400x, respectively.

Most compound microscopes provide two forms of lighting for the specimens viewed. *Incident lighting* (not pictured) is light coming from above the specimen. *Transmitted lighting* comes from below. The *diaphragm* or *aperture* controls the amount of light that passes through the specimen when transmitted lighting is used.

The *focus knob* is used to move the stage higher and lower to bring the specimen into clearer focus. Better quality microscopes have both a coarse focus (shown) and a fine focus (not pictured).

To perform an observation, follow these steps:

1. Set the microscope on a large, flat, steady surface. You'll want to avoid surfaces that are subject to movement or vibration and areas where the microscope might be bumped.

2. Turn on the bottom light and adjust the diaphragm or aperture to the widest opening.

3. Choose the lowest power objective lens by rotating the nosepiece until the objective rests over the hole in the center of the stage and you hear it click into place.

4. Place a slide on the stage, under the stage clips, with the specimen centered.

5. Carefully raise the stage to the highest level without *touching the objective lens*. It is possible to break the slide and cover slip and to damage the objective lens if you raise the stage too high. It is recommended that you look from the side while performing this step.

6. Adjust the focus knob until the image is clear. If you have both coarse and fine focus, adjust the coarse focus first, then refine with the fine focus knob.

7. Adjust the diaphragm to progressively lower light settings until you get the clearest image with sharpest contrast.

8. Move the slide around to get an overview of the specimen. Center any area you want to examine further. Create any sketches you need for 40x magnification.

9. Next, choose the 10x objective lens by rotating the nosepiece, being careful not to crush the slide or cover slip. Refocus for 100x magnification. More light is required for higher magnifications, so you will need to adjust the diaphragm for a larger opening until you get the clearest image.

10. Make any necessary sketches of the image at 100x magnification.

11. Repeat the steps for the 40x objective lens to get 400x magnification, again making any necessary drawings.

12. Remember to store your microscope and permanent mount slides appropriately. Clean any temporary slides and put them safely away.

PREPARING YOUR OWN SLIDES

There are several types of mounts and staining methods you will use when making your own slides. A *whole mount* is made when you wish to view an entire specimen like a small insect or a seed. If the specimen is relatively large, a concavity slide can be used for the mounting. Concavity slides have a small well in the center to allow for specimens that are too thick for an ordinary mount.

Fluids, like blood or cheek scrapings are often made into *smears*. To make a smear, place some of the fluid on a clean slide. Use the edge of a coverslip to spread the material into a thin layer on the slide. Smears are stained after being allowed to air dry. Do this by placing 1-2 drops of stain on the slide and leaving for 5 minutes. Then, carefully rinse under a small, slow stream of water and allow to dry before applying the coverslip.

Section mounts use very thin, nearly transparent slices of material. With a relatively solid specimen like onion skin you may choose to stain using a dip and rinse process. Place the specimen in a small amount of stain for about 3 minutes. Remove it with tweezers and dip it repeatedly in a small amount of water. You may then make a *wet mount* with your stained specimen. First, place a drop of water on a clean slide. Using tweezers, lay the stained specimen in the drop of water. Hold a coverslip at a 45-degree angle to the slide and carefully lower it onto the specimen to minimize air bubbles. Any remaining bubbles can be removed by tapping carefully with a pencil eraser.

For smaller and more delicate material, you can pull the stain. First wet mount the unstained specimen under a coverslip as above. Then put a drop of stain onto the slide on one side of the specimen. Place a piece of blotter paper or paper towel on the other side of the specimen close to the coverslip. The stain will be wicked toward the paper, thereby pulling it under the coverslip and across the specimen. You can even pull a second stain to better differentiate the specimen's structures.

The particular stain chosen for a mount depends on which of the specimen's features you wish to accentuate. Eosin tints a pink color. It stains alkalis more thoroughly than acids and therefore gives a deeper color to the cytoplasm of cells. Methylene blue shades acids darker than alkalis and causes the nuclei to show more clearly. Iodine stains all complex carbohydrates, but the color achieved depends on the type of carbohydrate. It will accentuate starch grains in plant cells with a dark blue color. The glycogen stores in animal cells will show as red.

LESSONS

LIFE IN GENERAL

LESSON 01: THE SCIENTIFIC METHOD

READING & RESEARCH

- ☐ Read "The Scientific Method: 'Doing' Science" in the *Fundamentals* section of this study guide.

- ☐ Read USE "Research on the Web", pp. 394-395.

- ☐ Read DSE "How Scientists Work", pp. 14-15.

DEFINITIONS

- ☐ Define the following terms, noting any Latin or Greek word parts, and place in the Definitions section of your Science Notebook:

    o *scientific method*

    o *hypothesis*

    o *independent variable*

    o *dependent variable*

LABS & ACTIVITIES

- ☐ Do Lab 01: *From Madness to Method* in the Labs section of this study guide.

- ☐ File Lab Report 01 in the Labs section of the Science Notebook.

REPORT

- ☐ Write 1 paragraph about what you've learned and file in the Research Reports section of your Science Notebook.

LIFE IN GENERAL

SCIENCE HISTORY

- ☐ Record the birth and death dates of Sir Francis Bacon as well as the publication date of his <u>Novum Organum</u>.

- ☐ Record the birth and death dates of Rene Descartes as well as the publication date of his <u>Discourse on the Method of Rightly Conducting One's Reason and of Seeking Truth in the Sciences</u>.

- ☐ Record any other significant dates you've encountered in the Dates section of your Science Notebook and on the timeline. Include discoveries, inventions and works, as well as the birth and death dates of important thinkers.

LIFE IN GENERAL

LESSON 02: CLASSIFYING LIFE

READING & RESEARCH

- [x] Read KSE "Classification of Living Things" p. 52-53.
- [x] Read USE "Classification" p. 340-343.
- [x] Read USE "Classifying Plants" p. 294-295.
- [ ] Read DSE "Classifying Living Things", pp. 310-311.

DEFINITIONS

- [ ] In DB, read the definitions in "The Classification of Living Things", p. 340.
- [ ] Define the following terms, noting any Latin or Greek word parts, and place in the Definitions section of your Science Notebook:
    - *taxonomy*
    - *species*
    - *binomial*

LABS & ACTIVITIES

- [ ] Do Lab 02: *Categorically Speaking* in the Labs section of this study guide.
- [ ] File Lab Report in the Labs section of the Science Notebook.

REPORT

- [ ] Write a 2-3 paragraph report on what you've learned and file in the Research Reports section of your Science Notebook.

© 2010 Classical Education Resources, LLC.   All rights reserved.

LIFE IN GENERAL

SCIENCE HISTORY

☐ Record the birth and death dates of Carolus Linnaeus as well as the publication of his classification system, *Systema Naturae,* in the Dates section of your Science Notebook and on the timeline.

☐ Record any other significant dates you've encountered in the Dates section of your Science Notebook and on the timeline. Include discoveries, inventions and works, as well as the birth and death dates of important thinkers.

MEMORY WORK

☐ Begin memorizing *List A: The Seven Major Groups of Classical Taxonomy.*

LESSON 03: LIFE AND ITS CYCLES

READING & RESEARCH

- [x] Read USE "Life Cycles", pp. 328-329.

- [ ] Read DSE "Living Things", p. 305.

- [ ] Read DSE "What is Life", p. 306.

DEFINITIONS

- [ ] In DB, read the definitions in "Life and Life Cycles", p. 236.

- [ ] Define the following terms, noting any Latin or Greek word parts, and place in the Definitions section of your Science Notebook:

    - *biology*, DB p. 230

    - *life cycle*, DB p. 236

    - *metamorphosis*, DB p. 277

LABS & ACTIVITIES

- [ ] Do Lab 03: *It's Alive!* in the Labs section of this study guide.

- [ ] File Lab Report in the Labs section of the Science Notebook.

REPORT

- [ ] Write a 1-2 paragraph report on what you've learned and file in the Research Reports section of your Science Notebook.

SCIENCE HISTORY

- [ ] Record the birth and death dates of Friedrich Wohler along with his contribution and its date.

- [ ] Record the birth and death dates of naturalist Henry Bates.

- [ ] Record any other significant dates you've encountered in the Dates section of your Science Notebook and on the timeline. Include discoveries, inventions and works, as well as the birth and death dates of important thinkers.

MEMORY WORK

- [ ] Continue memorizing *List A: The Seven Major Groups of Classical Taxonomy.*

LIFE IN GENERAL

LESSON 04: THE MICROSCOPIC WORLD

READING & RESEARCH

☐ Read "*Using the Microscope*" in the *Fundamentals* section of this study guide.

☑ Read KSE "Microscopes", pp. 268-269.

☑ Read USE "Optical Instruments" the section titled "Optical Microscopes" p. 222.

☐ Read DSE "Optical Instruments" the section titled "Compound Microscope" p. 198.

☐ Read the instructions that came with your microscope.

DEFINITIONS

☐ In DB, read the paragraph titled "Optical Instruments: Microscope", p. 54.

☐ Define the following terms, noting any Latin or Greek word parts, and place in the Definitions section of your Science Notebook:

  o *microscopy*

  o *microbiology*

LABS & ACTIVITIES

☐ Do Lab 04: *Ready for My Close-up* in the Labs section of this study guide.

☐ File Lab Report forms in the Labs section of your Science Notebook.

☐ Label Diagram 04-1, *Compound Microscope,* and file in the Diagrams section of your Science Notebook.

REPORT

☐ Write a 1 paragraph report on what you've learned and file in the Research Reports section of your Science Notebook.

© 2010 Classical Education Resources, LLC.   All rights reserved.

LIFE IN GENERAL

SCIENCE HISTORY

☐ Record the birth and death dates of Antonie van Leeuwenhoek along with his major contribution to science.

☐ Record the birth and death dates of Zacharias Janssen along with his invention and its approximate date.

☐ Record the birth and death dates of Robert Hooke along with the publication date of his <u>Micrographia</u>.

☐ Record the birth and death dates of Louis Pasteur along with his contribution and its date.

☐ Record the birth and death dates of Vladimir Zworykin along with his invention and its approximate date.

☐ Record any other significant dates you've encountered in the Dates section of your Science Notebook and on the timeline. Include discoveries, inventions and works, as well as the birth and death dates of important thinkers.

MEMORY WORK

☐ Continue memorizing *List A: The Seven Major Groups of Classical Taxonomy.*

LIFE IN GENERAL

LESSON 05: CELLS: THE BUILDING BLOCKS OF LIFE

READING & RESEARCH

- [ ] Read USE "Plant Cells", pp. 250-251.

- [ ] Read USE "Animal Cells", pp. 298-299.

- [ ] Read DSE "Cells", pp. 338-339.

DEFINITIONS

- [ ] In DB, read the definitions in "The Structure of Living Things" up to Cell Division, pp. 238-240.

- [ ] Define the following terms, noting any Latin or Greek word parts, and place in the Definitions section of your Science Notebook:

    - *organism*
    - *cell*
    - *nucleus*

LABS & ACTIVITIES

- [ ] Do Lab 05: *Same Difference* in the Labs section of this study guide.

- [ ] File Lab Report forms in the Labs section of your Science Notebook.

- [ ] Label Diagram 05-1, *Plant Cell,* and *Diagram 05-2, Animal Cell,* and file in the Diagrams section of your Science Notebook.

REPORT

- [ ] Write a 2-3 paragraph report on what you've learned and file in the Research Reports section of your Science Notebook.

© 2010 Classical Education Resources, LLC.   All rights reserved.

LIFE IN GENERAL

SCIENCE HISTORY

- ☐ Record the birth and death dates of Theodor Schwann and Jakob Mathias along with their contribution.

- ☐ Record the contribution of Edouard Chatton.

- ☐ Record any other significant dates you've encountered in the Dates section of your Science Notebook and on the timeline. Include discoveries, inventions and works, as well as the birth and death dates of important thinkers.

MEMORY WORK

- ☐ Continue memorizing *List A: The Seven Major Groups of Classical Taxonomy*.

LESSON 06: REPRODUCTION

READING & RESEARCH

☐ Read USE "Creating New Life", pp. 324-325.

☐ Read DSE "Asexual Reproduction", p. 366.

DEFINITIONS

☐ In DB, read the definitions in "Types of Reproduction", pp. 320-321.

☐ Define the following terms, noting any Latin or Greek word parts, and place in the Definitions section of your Science Notebook:

- *reproduction*
- *gamete*
- *zygote*

LABS & ACTIVITIES

☐ Do Lab 06: *Hydra Hijinx and Future Frogs* in the Labs section of this study guide.

☐ File Lab Report forms in the Labs section of your Science Notebook.

REPORT

☐ Write a 2-3 paragraph report on what you've learned and file in the Research Reports section of your Science Notebook.

SCIENCE HISTORY

☐ Record the date of van Leeuwenhoek's observation of hydra budding in the Dates section of your Science Notebook and on the timeline.

☐ Record any other significant dates you've encountered in the Dates section of your Science Notebook and on the timeline. Include discoveries, inventions and works, as well as the birth and death dates of important thinkers.

MEMORY WORK

☐ Continue memorizing *List A: The Seven Major Groups of Classical Taxonomy.*

LIFE IN GENERAL

LESSON 07: CELL DIVISION

READING & RESEARCH

☐ Read KSE "Animal Reproduction", pp. 88-89.

☐ Read DSE "Sexual Reproduction", p. 367.

DEFINITIONS

☐ In DB, read the definitions in "The Structure of Living Things: Cell Division", pp. 240-241.

☐ In DB, read the definitions in "Cell Division for Reproduction", pp. 322-323.

☐ Define the following terms, noting any Latin or Greek word parts, and place in the Definitions section of your Science Notebook:

- *mitosis*
- *meiosis*

LABS & ACTIVITIES

☐ Do Lab 07: *Split Decision* in the Labs section of this study guide.

☐ File Lab Report forms in the Labs section of your Science Notebook.

☐ Label Diagram 07-1, *Mitosis*, Diagram 07-2, *Meiosis I*, and Diagram 07-3, *Meiosis II*, and file in the Diagrams section of your Science Notebook.

REPORT

☐ Write a 2-3 paragraph report on what you've learned and file in the Research Reports section of your Science Notebook.

LIFE IN GENERAL

SCIENCE HISTORY

☐ Record the date of the first adult mammal clone.

☐ Record any other significant dates you've encountered in the Dates section of your Science Notebook and on the timeline. Include discoveries, inventions and works, as well as the birth and death dates of important thinkers.

MEMORY WORK

☐ Continue memorizing *List A: The Seven Major Groups of Classical Taxonomy.*

LESSON 08: GENETICS

READING & RESEARCH

- ☐ Read KSE "Genes and Chromosomes", p. 135.
- ☐ Read USE "Genetics", pp. 380-381.
- ☐ Read DSE "Genetics", pp. 364-365.

DEFINITIONS

- ☐ In DB, read the definitions in "Genetics and Heredity", pp. 324-326.
- ☐ Define the following terms, noting any Latin or Greek word parts, and place in the Definitions section of your Science Notebook:
    - *gene*
    - *genetics*
    - *chromosomes*

LABS & ACTIVITIES

- ☐ Do Lab 08: *A New Pair of Genes* in the Labs section of this study guide.
- ☐ File Lab Report in Labs section of the Science Notebook.

REPORT

- ☐ Write a 2-3 paragraph report on what you've learned and file in the Research Reports section of your Science Notebook.

SCIENCE HISTORY

- ☐ Record the birth and death dates of Gregor Mendel, along with his contribution.

LIFE IN GENERAL

- ☐ Record the birth date of James Watson.

- ☐ Record the birth and death dates of Francis Crick.

- ☐ Record the birth and death dates of Maurice Wilkins. *Notice anything interesting?*

- ☐ Record the birth and death dates of Rosalind Franklin along with her contribution to science.

- ☐ Record the date of the discovery of the structure of the DNA molecule.

- ☐ Record the date of the Nobel Prize for Watson, Crick and Wilkins.

- ☐ Record the date scientists completed the sequencing of the human genome.

- ☐ Record any other significant dates you've encountered in the Dates section of your Science Notebook and on the timeline. Include discoveries, inventions and works, as well as the birth and death dates of important thinkers.

MEMORY WORK

- ☐ Recite *List A: The Seven Major Groups of Classical Taxonomy* for your family.

- ☐ File *List A* in the Memory Work section of your Science Notebook.

LESSON 09: SIMPLE ORGANISMS

READING & RESEARCH

☐ Read KSE "Single-Celled Organisms", p. 54.

☐ Read KSE "Bacteria and Viruses", p. 136.

☐ Read DSE "Single-celled Organisms", p. 314.

☐ Read DSE "Bacteria", p. 313.

☐ Read DSE "Viruses", p. 312.

DEFINITIONS

☐ Define the following terms, noting any Latin or Greek word parts, and place in the Definitions section of your Science Notebook:

- *moneran*
- *protist*
- *bacteria*
- *virus*

LABS & ACTIVITIES

☐ Do Lab 09: It's a Small World After All – BEK experiment 39, pp. 86-87.

➤ This is the first lab that does not have the objective, materials and procedure written out for you. You should use copies of the Lab Report forms to create a complete Lab Report on your own. Be sure to include the Objective & Materials form and the Procedure form. You may use either a Results, Narrative form to write out your results or a Results, Observation form to draw them. When finished, file the complete Lab Report in the Labs section of your Science Notebook.

➢ You may substitute clear disposable cups for small clear glasses.

REPORT

☐ Write a 2-3 paragraph report on what you've learned and file in the Research Reports section of your Science Notebook.

SCIENCE HISTORY

☐ Record the birth and death dates of Robert Koch along with his contribution mentioned in this lesson.

☐ Record any other significant dates you've encountered in the Dates section of your Science Notebook and on the timeline. Include discoveries, inventions and works, as well as the birth and death dates of important thinkers.

MEMORY WORK

☐ Begin memorizing *List B: The Five Kingdoms of Classical Taxonomy.*

LIFE IN GENERAL

LESSON 10: NEITHER PLANT NOR ANIMAL

READING & RESEARCH

- ☐ Read KSE "Fungi and Lichens", p. 55.
- ☐ Read USE "Fungi", pp. 284-285.
- ☐ Read DSE "Fungi", p. 315.

DEFINITIONS

- ☐ Define the following terms, noting any Latin or Greek word parts, and place in the Definitions section of your Science Notebook:
    - *fungi*
    - *lichen*
    - *chlorophyll*
    - *symbiosis*

LABS & ACTIVITIES

- ☐ Do Lab 10: *Fungus Among Us* in the Labs section of this study guide.
- ☐ File the Lab Report in the Labs section of the Science Notebook.

REPORT

- ☐ Write a 2-3 paragraph report on what you've learned and file in the Research Reports section of your Science Notebook.

SCIENCE HISTORY

☐ Record the birth and death dates of Sir Alexander Fleming, along with the date of his serendipitous discovery.

☐ Record any other significant dates you've encountered in the Dates section of your Science Notebook and on the timeline. Include discoveries, inventions and works, as well as the birth and death dates of important thinkers.

MEMORY WORK

☐ Continue memorizing *List B: The Five Kingdoms of Classical Taxonomy*.

LESSON 11: STRUCTURE OF PLANTS

READING & RESEARCH

- ☐ Read KSE "Plant Anatomy", pp. 56-57.
- ☐ Read USE "Stems and Roots", pp. 252-253.
- ☐ Read USE "Plant Tissue", pp. 254-255.
- ☐ Read USE "Inside Older Plants", pp. 256-257.
- ☐ Read DSE "Transport in Plants", p. 341.

DEFINITIONS

- ☐ In DB, read the definitions in "Vascular Plants", pp. 242-243.
- ☐ In DB, read the definitions in "Stems and Roots", pp. 244-245.
- ☐ In DB, read the definitions in "Plant Fluid Transportation", pp. 252-253.
- ☐ In DB, read the definitions in "Inside an Older Plant", p. 246.
- ☐ Define the following terms, noting any Latin or Greek word parts, and place in the Definitions section of your Science Notebook:
    - *monocotyledon*
    - *dicotyledon*

LABS & ACTIVITIES

- ☐ Do Lab 11: The Incredible Drinking Plant – BEK experiment 6, pp. 18-19.
    - ➢ You will need a white carnation with a long stem for this lab activity.
    - ➢ You can use clear disposable cups instead of the two glasses.
    - ➢ You may need to brace the flower during the 48 hours of this lab experiment.

BOTANY – THE STUDY OF PLANTS

> ➢ You can substitute a celery stalk, but the carnation gives a more dramatic effect.

☐ Write a complete Lab Report and file in the Labs section of the Science Notebook.

REPORT

☐ Write a 2-3 paragraph report on what you've learned and file in the Research Reports section of your Science Notebook.

SCIENCE HISTORY

☐ Research the birth and death date of Joseph Priestly, along with his contribution to botany.

☐ Record any other significant dates you've encountered in the Dates section of your Science Notebook and on the timeline. Include discoveries, inventions and works, as well as the birth and death dates of important thinkers.

MEMORY WORK

☐ Continue memorizing *List B: The Five Kingdoms of Classical Taxonomy*.

LESSON 12: PLANT LEAVES AND FOOD PRODUCTION

READING & RESEARCH

- [ ] Read USE "Leaves", pp. 258-259.
- [ ] Read USE "Leaf structure", pp. 260-261.
- [ ] Read USE "Plant Food", pp. 264-267.
- [ ] Read DSE "Photosynthesis", p. 340.

DEFINITIONS

- [ ] In DB, read the definitions in "Leaves", pp. 248-250.
- [ ] In DB, read the definitions in "Plant Food Production", pp. 254-255.
- [ ] Define the following terms, noting any Latin or Greek word parts, and place in the Definitions section of your Science Notebook:
    - *photosynthesis*
    - *chloroplast*

LABS & ACTIVITIES

- [ ] Do Lab 12: Going Green – BEK experiment 26, pp. 58-59.
    - ➤ You will need a mustard plant from the Botanical Discoveries Kit or another houseplant. You should plant it 2 – 3 weeks before doing this experiment.
    - ➤ Use masking tape, and tape all around the edges of the construction paper so that no light gets through.
- [ ] Write a complete Lab Report and file in the Labs section of the Science Notebook.

BOTANY – THE STUDY OF PLANTS

REPORT

☐ Write a 2-3 paragraph report on what you've learned and file in the Research Reports section of your Science Notebook.

SCIENCE HISTORY

☐ Record the birth and death date of Jan Ingenhousz, along with his discovery.

☐ Record any other significant dates you've encountered in the Dates section of your Science Notebook and on the timeline. Include discoveries, inventions and works, as well as the birth and death dates of important thinkers.

MEMORY WORK

☐ Continue memorizing *List B: The Five Kingdoms of Classical Taxonomy.*

BOTANY – THE STUDY OF PLANTS

LESSON 13: PLANT SENSITIVITY

READING & RESEARCH

☐ Read USE "Plant Sensitivity", pp. 268 – 269.

DEFINITIONS

☐ In DB, read the definitions in "Plant Sensitivity", p. 251.

☐ Define the following terms, noting any Latin or Greek word parts, and place in the Definitions section of your Science Notebook:

  o *tropisim*

LABS & ACTIVITIES

☐ Do Lab 13: Into the Light – BEK experiment 23, pp. 52-53.

> ➢ You will need a freshly sprouted mustard plant from the Botanical Discoveries Kit or freshly sprouted pinto beans. Either requires about a week of lead time.

> ➢ Use a germination cup and germination mixture from the Botanical Discoveries Kit.

☐ Write a complete Lab Report and file in the Labs section of the Science Notebook.

REPORT

☐ Write a 2-3 paragraph report on what you've learned and file in the Research Reports section of your Science Notebook.

SCIENCE HISTORY

☐ Record any significant dates you've encountered in the Dates section of your Science Notebook and on the timeline. Include discoveries, inventions and works, as well as the birth and death dates of important thinkers.

© 2010 Classical Education Resources, LLC.   All rights reserved.

MEMORY WORK

☐ Recite *List B: The Five Kingdoms of Classical Taxonomy* for your family.

☐ File *List B* in the Memory Work section of your Science Notebook.

BOTANY - THE STUDY OF PLANTS

LESSON 14: FLOWERS

READING & RESEARCH

☐ Read KSE "Flowering Plants", pp. 59-61.

☐ Read USE "Flowering Plants", pp. 270-273.

☐ Read DSE "Flowering Plants", pp. 318-319.

DEFINITIONS

☐ In DB, read the definitions in "Flowers", pp. 256-257.

☐ In DB, read the definitions in "Reproduction in a Flowering Plant", pp. 258-259.

☐ Define the following terms, noting any Latin or Greek word parts, and place in the Definitions section of your Science Notebook:

- *pollen*
- *ovules*

LABS & ACTIVITIES

☐ Do Lab 14: *What Cot?* in the Labs section of this study guide.

☐ File Lab Report forms in the Labs section of your Science Notebook.

REPORT

☐ Write a 2-3 paragraph report on what you've learned and file in the Research Reports section of your Science Notebook.

© 2010 Classical Education Resources, LLC.   All rights reserved.

BOTANY - THE STUDY OF PLANTS

SCIENCE HISTORY

☐ Record any significant dates you've encountered in the Dates section of your Science Notebook and on the timeline. Include discoveries, inventions and works, as well as the birth and death dates of important thinkers.

MEMORY WORK

☐ Begin memorizing *List C: Parts of a Flower*.

BOTANY - THE STUDY OF PLANTS

LESSON 15: FRUITS AND SEEDS

READING & RESEARCH

- [ ] Read KSE "Fruits and Seeds", pp. 62-63.

- [ ] Read USE "Seeds and Fruit", pp. 274-277.

DEFINITIONS

- [ ] In DB, read the definitions in "Seeds and Germination", pp. 260-261.

- [ ] In DB, read the definitions in "Fruit", p. 262.

- [ ] Define the following terms, noting any Latin or Greek word parts, and place in the Definitions section of your Science Notebook:

    - *fertilization*
    - *ovary*
    - *embryo*
    - *germination*

LABS & ACTIVITIES

- [ ] Do Lab 15: Beanie Babies – BEK experiment 18, pp. 42-43.

    > If you have the plastic gallon jar it may be used in lieu of a quart jar.

- [ ] Write a complete Lab Report and file in the Labs section of the Science Notebook.

REPORT

- [ ] Write a 2-3 paragraph report on what you've learned and file in the Research Reports section of your Science Notebook.

© 2010 Classical Education Resources, LLC.   All rights reserved.

BOTANY - THE STUDY OF PLANTS

SCIENCE HISTORY

☐ Research the life and work of John Chapman and record significant dates. Be sure to record his better known nickname.

☐ Record any other significant dates you've encountered in the Dates section of your Science Notebook and on the timeline. Include discoveries, inventions and works, as well as the birth and death dates of important thinkers.

MEMORY WORK

☐ Continue memorizing *List C: Parts of a Flower.*

BOTANY - THE STUDY OF PLANTS

LESSON 16: VEGETATIVE REPRODUCTION

READING & RESEARCH

☐ Read KSE "Nonflowering Plants", p. 58.

☐ Read USE "New Plants from Old", pp. 278-279.

☐ Read USE "Flowerless Plants", pp. 282-283.

☐ Read DSE "Plants without Flowers", p. 316.

DEFINITIONS

☐ In DB, read the definitions in "Vegetative Reproduction", p. 263.

☐ Define the following terms, noting any Latin or Greek word parts, and place in the Definitions section of your Science Notebook:

- *propagation*

LABS & ACTIVITIES

☐ Do Lab 16: The Eyes Have It – BEK experiment 29, pp. 64-65.

> You will need a potato with eyes. It is suggested you store about four potatoes in a dark place for a few weeks before you plan to do the experiment.

> If you have the plastic gallon jar it may be used in lieu of a quart jar.

☐ Write a complete Lab Report and file in the Labs section of the Science Notebook.

REPORT

☐ Write a 2-3 paragraph report on what you've learned and file in the Research Reports section of your Science Notebook.

SCIENCE HISTORY

☐ Record any significant dates you've encountered in the Dates section of your Science Notebook and on the timeline. Include discoveries, inventions and works, as well as the birth and death dates of important thinkers.

MEMORY WORK

☐ Recite *List C: Parts of a Flower.*

☐ File *List C* in the Memory Work section of your Science Notebook.

LESSON 17: MOLLUSKS, CNIDARIA & ECHINODERMS

READING & RESEARCH

- ☐ Read KSE "Marine Invertebrates", p. 70.
- ☐ Read KSE "Mollusks", p. 71.
- ☐ Read DSE "Mollusks", p. 324.
- ☐ Read DSE "Jellyfish, Anemones, and Corals", p. 320.
- ☐ Read DSE "Starfish and Sea Squirts", p. 325.

DEFINITIONS

- ☐ Define the following terms, noting any Latin or Greek word parts, and place in the Definitions section of your Science Notebook:
    - o *invertebrate*
    - o *mollusk*
    - o *echinoderm*
    - o *cnidarian*

LABS & ACTIVITIES

- ☐ Do Lab 17: Sticky Situation – BEK experiment 54, pp. 116-117.
    - ➢ Try creating a sea creature! You'll need up to eight suction cups instead of just one and a few feet of yarn or string. Have fun!
- ☐ Write a complete Lab Report and file in the Labs section of the Science Notebook.

ZOOLOGY - THE STUDY OF ANIMALS

REPORT

☐ Write a 2-3 paragraph report on what you've learned and file in the Research Reports section of your Science Notebook.

SCIENCE HISTORY

☐ Research Jacques-Yves Cousteau and record the dates of his birth and death.

☐ Record any significant dates you've encountered in the Dates section of your Science Notebook and on the timeline. Include discoveries, inventions and works, as well as the birth and death dates of important thinkers.

MEMORY WORK

☐ Begin memorizing *List D: Animal Kingdom Phyla*.

ZOOLOGY – THE STUDY OF ANIMALS

LESSON 18: WORMS

READING & RESEARCH

☐ Read KSE "Worms", p. 72.

☐ Read DSE "Worms", p. 321.

DEFINITIONS

☐ Define the following terms, noting any Latin or Greek word parts, and place in the Definitions section of your Science Notebook:

- *parasite*
- *segment*

LABS & ACTIVITIES

☐ Do Lab 18: Sensitive Worms – BEK experiment 55, pp. 118-119.

> You'll need approximately four to six earthworms for this activity.

☐ Write a complete Lab Report and file in the Labs section of the Science Notebook.

REPORT

☐ Write a 2-3 paragraph report on what you've learned and file in the Research Reports section of your Science Notebook.

SCIENCE HISTORY

☐ Research the latest date in which leeches were used in medicine. Surprised? Record the approximate date.

ZOOLOGY – THE STUDY OF ANIMALS

☐ Record any significant dates you've encountered in the Dates section of your Science Notebook and on the timeline. Include discoveries, inventions and works, as well as the birth and death dates of important thinkers.

MEMORY WORK

☐ Continue memorizing *List D: Animal Kingdom Phyla.*

LESSON 19: ARTHROPODS: ARACHNIDS AND CRUSTACEANS

READING & RESEARCH

☐ Read KSE "Crustaceans", p. 73.

☐ Read KSE "Spiders, Centipedes, and Scorpions", p. 74.

☐ Read DSE "Arthropods", first page only, p. 322.

DEFINITIONS

☐ Define the following terms, noting any Latin or Greek word parts, and place in the Definitions section of your Science Notebook:

- *arthropod*
- *arachnid*
- *crustacean*

LABS & ACTIVITIES

☐ For Lab 19: *Study a Spider*, read and explore the book Uncover a Tarantula. Using a Lab Report Conclusion form, write a few sentences about each section of the book detailing the most interesting facts you read. File in the Labs section of your Science Notebook.

☐ Label Diagram 19-1, *Tarantula,* and file in the Diagrams section of your Science Notebook.

REPORT

☐ Write a 2-3 paragraph report on what you've learned and file in the Research Reports section of your Science Notebook.

SCIENCE HISTORY

☐ Record any significant dates you've encountered in the Dates section of your Science Notebook and on the timeline. Include discoveries, inventions and works, as well as the birth and death dates of important thinkers.

MEMORY WORK

☐ Continue memorizing *List D: Animal Kingdom Phyla.*

LESSON 20: ARTHROPODS: INSECTS

READING & RESEARCH

☐ Read KSE "Insects", pp. 75-77.

☐ Read DSE "Arthropods", second page only, p. 323.

DEFINITIONS

☐ Define the following terms, noting any Latin or Greek word parts, and place in the Definitions section of your Science Notebook:

- *insect*
- *thorax*

LABS & ACTIVITIES

☐ Do Lab 20: *Footloose* in the Labs section of this study guide.

☐ File Lab Report forms in the Labs section of your Science Notebook.

REPORT

☐ Write a 2-3 paragraph report on what you've learned and file in the Research Reports section of your Science Notebook.

SCIENCE HISTORY

☐ Record the birth and death dates of entomologist Jean-Henri Fabre.

☐ Record any other significant dates you've encountered in the Dates section of your Science Notebook and on the timeline. Include discoveries, inventions and works, as well as the birth and death dates of important thinkers.

MEMORY WORK

☐ Continue memorizing *List D: Animal Kingdom Phyla.*

LESSON 21: FISH

READING & RESEARCH

- ☐ Read KSE "Fish", pp. 78-79.
- ☐ Read DSE "Fish", pp. 326-327.

DEFINITIONS

- ☐ Define the following terms, noting any Latin or Greek word parts, and place in the Definitions section of your Science Notebook:
    - *vertebrate*
    - *fish*
    - *cartilage*

LABS & ACTIVITIES

- ☐ Do Lab 21: Fishy Birthdays in the Labs section of this study guide.
- ☐ File the Lab Report in the Labs section of the Science Notebook.

REPORT

- ☐ Write a 2-3 paragraph report on what you've learned and file in the Research Reports section of your Science Notebook.

SCIENCE HISTORY

- ☐ Research the fish *coelacanth* and record the date of that scientific surprise.
- ☐ Record any significant dates you've encountered in the Dates section of your Science Notebook and on the timeline. Include discoveries, inventions and works, as well as the birth and death dates of important thinkers.

MEMORY WORK

☐ Continue memorizing *List D: Animal Kingdom Phyla.*

ZOOLOGY – THE STUDY OF ANIMALS

LESSON 22: AMPHIBIANS

READING & RESEARCH

☐ Read KSE "Amphibians", pp. 80-81.

☐ Read DSE "Amphibians", pp. 328-329.

DEFINITIONS

☐ Define the following terms, noting any Latin or Greek word parts, and place in the Definitions section of your Science Notebook:

- *amphibian*
- *cold-blooded*

LABS & ACTIVITIES

☐ For Lab 22: *Focus on a Frog,* read and explore the book Uncover a Frog. Using a Lab Report Conclusion form, write a few sentences about each section of the book detailing the most interesting facts you read. File in the Labs section of your Science Notebook.

☐ Label Diagram 22-1, *External Anatomy of the Frog,* Diagram 22-2, *Internal Anatomy of the Frog,* and Diagram 22-3, *Frog Life Cycle,* and file in the Diagrams section of your Science Notebook.

REPORT

☐ Write a 2-3 paragraph report on what you've learned and file in the Research Reports section of your Science Notebook.

© 2010 Classical Education Resources, LLC.   All rights reserved.

SCIENCE HISTORY

☐ Record any significant dates you've encountered in the Dates section of your Science Notebook and on the timeline. Include discoveries, inventions and works, as well as the birth and death dates of important thinkers.

MEMORY WORK

☐ Continue memorizing *List D: Animal Kingdom Phyla.*

ZOOLOGY – THE STUDY OF ANIMALS

LESSON 23: REPTILES

READING & RESEARCH

☐ Read KSE "Reptiles", pp. 82-83.

☐ Read DSE "Reptiles", pp. 330-331.

DEFINITIONS

☐ Define the following terms, noting any Latin or Greek word parts, and place in the Definitions section of your Science Notebook:

- *reptile*
- *poikilothermic*

LABS & ACTIVITIES

☐ For Lab 23: *Consider a Cobra*, read and explore the book <u>Uncover a Cobra</u>. Using a Lab Report Conclusion form, write a few sentences about each section of the book detailing the most interesting facts you read. File in the Labs section of your Science Notebook.

☐ Label Diagram 23-1, *King Cobra*, and file in the Diagrams section of your Science Notebook.

REPORT

☐ Write a 2-3 paragraph report on what you've learned and file in the Research Reports section of your Science Notebook.

SCIENCE HISTORY

☐ Record any significant dates you've encountered in the Dates section of your Science Notebook and on the timeline. Include discoveries, inventions and works, as well as the birth and death dates of important thinkers.

© 2010 Classical Education Resources, LLC.   All rights reserved.

MEMORY WORK

☐ Continue memorizing *List D: Animal Kingdom Phyla.*

ZOOLOGY – THE STUDY OF ANIMALS

LESSON 24: BIRDS

READING & RESEARCH

☐ Read KSE "Birds", pp. 84-85.

☐ Read DSE "Birds", pp. 332-333.

DEFINITIONS

☐ Define the following terms, noting any Latin or Greek word parts, and place in the Definitions section of your Science Notebook:

- *bird*
- *warm-blooded*

LABS & ACTIVITIES

☐ Do Lab 24: Whooo Eats What? in the Labs section of this study guide.

➢ One or two disposable dissecting pans are a convenient substitution for paper plates due to their larger size and flat design.

☐ File Lab Report including the dissection worksheet in the Labs section of the Science Notebook.

REPORT

☐ Write a 2-3 paragraph report on what you've learned and file in the Research Reports section of your Science Notebook.

SCIENCE HISTORY

☐ Research the life and work of John James Audubon and record any important dates you encounter.

© 2010 Classical Education Resources, LLC.   All rights reserved.

☐ Record any other significant dates you've encountered in the Dates section of your Science Notebook and on the timeline. Include discoveries, inventions and works, as well as the birth and death dates of important thinkers.

MEMORY WORK

☐ Continue memorizing *List D: Animal Kingdom Phyla.*

LESSON 25: MAMMALS

READING & RESEARCH

☐ Read KSE "Mammals", pp. 86-87.

☐ Read DSE "Mammals", pp. 334-335.

DEFINITIONS

☐ Define the following terms, noting any Latin or Greek word parts, and place in the Definitions section of your Science Notebook:

- *mammal*
- *homiothermic*

LABS & ACTIVITIES

☐ Do Lab 25: *Hairy Situation* in the Labs section of this study guide.

☐ File Lab Report forms in the Labs section of your Science Notebook.

REPORT

☐ Write a 2-3 paragraph report on what you've learned and file in the Research Reports section of your Science Notebook.

SCIENCE HISTORY

☐ Research the life and contributions of Jane Goodall and record the dates.

☐ Record any other significant dates you've encountered in the Dates section of your Science Notebook and on the timeline. Include discoveries, inventions and works, as well as the birth and death dates of important thinkers.

MEMORY WORK

☐ Recite *List D: Animal Kingdom Phyla* for your family.

☐ File *List D* in the Memory Work section of your Science Notebook.

LESSON 26: RESPIRATORY SYSTEM

READING & RESEARCH

- [ ] Read KSE "Lungs and Breathing", pp. 124-125.

- [ ] Read USE "The Respiratory System", pp. 358-359.

DEFINITIONS

- [ ] In DB, read the definitions in "The Respiratory System", pp. 298-299.

- [ ] Define the following terms, noting any Latin or Greek word parts, and place in the Definitions section of your Science Notebook:

    o *respiration*

    o *ventilation*

LABS & ACTIVITIES

- [ ] Do Lab 26: Catch Your Breath – BEK experiment 88, pp. 186-187.

  > If you store your science supplies in a clear plastic tub, it can be temporarily emptied and used in lieu of the dishpan for this experiment.

  > If you have the plastic gallon jar it may be used in lieu of a quart jar.

  > Any similar sized vinyl tubing can be used in place of aquarium tubing.

- [ ] Write a complete Lab Report and file in the Labs section of the Science Notebook.

REPORT

- [ ] Write a 2-3 paragraph report on what you've learned and file in the Research Reports section of your Science Notebook.

SCIENCE HISTORY

☐ Record the birth and death dates of Andreas Vesalius, along with the publication date of his work The Structure of the Human Body. *See DSE p. 337*

☐ Record any other significant dates you've encountered in the Dates section of your Science Notebook and on the timeline. Include discoveries, inventions and works, as well as the birth and death dates of important thinkers.

MEMORY WORK

☐ Begin memorizing *List E: Major Systems of the Human Body.*

HUMAN ANATOMY

LESSON 27: CIRCULATORY SYSTEM

READING & RESEARCH

☐ Read KSE "The Heart and Circulation", pp. 120-121.

☐ Read KSE "Blood", p. 122.

☐ Read USE "The Circulatory System", pp. 350-351.

DEFINITIONS

☐ In DB, read the definitions in "Blood", pp. 286-287.

☐ In DB, read the definitions in "The Circulatory System", pp. 288-289.

☐ In DB, read the definitions in "The Heart", pp. 290-291.

☐ Define the following terms, noting any Latin or Greek word parts, and place in the Definitions section of your Science Notebook:

- erythrocyte
- leukocyte (leucocyte)
- artery
- vein
- capillary
- atria
- ventricle

LABS & ACTIVITIES

☐ Do Lab 27: What's Your Type? in the Labs section of this study guide.

☐ File the Lab Report in the Labs section of the Science Notebook.

REPORT

☐ Write a 2-3 paragraph report on what you've learned and file in the Research Reports section of your Science Notebook.

SCIENCE HISTORY

☐ Record the birth and death dates of William Harvey along with his important contribution.

☐ Record any other significant dates you've encountered in the Dates section of your Science Notebook and on the timeline. Include discoveries, inventions and works, as well as the birth and death dates of important thinkers.

MEMORY WORK

☐ Continue memorizing *List E: Major Systems of the Human Body*.

HUMAN ANATOMY

LESSON 28: DIGESTIVE SYSTEM

READING & RESEARCH

☐ Read KSE "Digestion", pp. 128-129.

☐ Read KSE "Waste Disposal", p. 131.

☐ Read USE "Digestion", pp. 354-355.

DEFINITIONS

☐ In DB, read the definitions in "The Digestive System", pp. 294-295.

☐ In DB, read the definitions in "The Urinary System", pp. 300-301.

☐ Define the following terms, noting any Latin or Greek word parts, and place in the Definitions section of your Science Notebook:

- *ingestion*

- *excretion*

LABS & ACTIVITIES

☐ Do Lab 28: Clean Up Your Act – BEK experiment 95, pp. 200-201.

➢ Cone-type coffee filters work best in funnels.

➢ If you have the plastic gallon jar it may be used in lieu of a quart jar.

➢ A clear disposable cup can be used in place of a small drinking glass.

➢ You can use the pipette from the Blood Typing Kit or the Microscope Slide Making Kit.

☐ Write a complete Lab Report and file in the Labs section of the Science Notebook.

© 2010 Classical Education Resources, LLC.   All rights reserved.

REPORT

- [ ] Write a 2-3 paragraph report on what you've learned and file in the Research Reports section of your Science Notebook.

SCIENCE HISTORY

- [ ] Record any significant dates you've encountered in the Dates section of your Science Notebook and on the timeline. Include discoveries, inventions and works, as well as the birth and death dates of important thinkers.

MEMORY WORK

- [ ] Continue memorizing *List E: Major Systems of the Human Body*.

HUMAN ANATOMY

LESSON 29: MUSCULAR SYSTEM

READING & RESEARCH

- [ ] Read KSE "Muscles and Movement", pp. 106-107.
- [ ] Read USE "Muscles", pp. 348-349.
- [ ] Read DSE "Muscles", p. 355.

DEFINITIONS

- [ ] In DB, read the definitions in "Muscles", pp. 282-283.
- [ ] Define the following terms, noting any Latin or Greek word parts, and place in the Definitions section of your Science Notebook:
    - *muscle*
    - *voluntary*
    - *involuntary*

LABS & ACTIVITIES

- [ ] Do Lab 29: *Don't Be Pushy* in the Labs section of this study guide.
- [ ] File Lab Report forms in the Labs section of your Science Notebook.

REPORT

- [ ] Write a 2-3 paragraph report on what you've learned and file in the Research Reports section of your Science Notebook.

© 2010 Classical Education Resources, LLC.   All rights reserved.

SCIENCE HISTORY

☐ Record the birth and death dates of Luigi Galvani along with his accidental contribution.

☐ Record any other significant dates you've encountered in the Dates section of your Science Notebook and on the timeline. Include discoveries, inventions and works, as well as the birth and death dates of important thinkers.

MEMORY WORK

☐ Recite *List E: Major Systems of the Human Body* for your family.

☐ File *List E* in the Memory Work section of your Science Notebook.

LESSON 30: SKELETAL SYSTEM

READING & RESEARCH

☐ Read KSE "The Skeleton", pp. 102-103.

☐ Read KSE "Bones and Joints", pp. 104-105.

☐ Read USE "Skeleton", pp. 346-347.

DEFINITIONS

☐ In DB, read the definitions in "The Skeleton", pp. 278-279.

☐ In DB, read the definitions in "Joints and Bone", pp. 280-281.

☐ Define the following terms, noting any Latin or Greek word parts, and place in the Definitions section of your Science Notebook:

- *skeleton*
- *joint*
- *articulation*

LABS & ACTIVITIES

☐ Do Lab 30: It's Not That Hard – BEK experiment 87, pp. 184-185.

➢ You will need one thin *uncooked* bone, either a wing or wishbone.

➢ If you have the plastic gallon jar it may be used in lieu of a quart jar.

☐ Write a complete Lab Report and file in the Labs section of the Science Notebook.

REPORT

☐ Write a 2-3 paragraph report on what you've learned and file in the Research Reports section of your Science Notebook.

SCIENCE HISTORY

☐ Research the discovery of Wilhelm Roentgoen and record any significant dates you encounter in your research.

☐ Record any other significant dates you've encountered in the Dates section of your Science Notebook and on the timeline. Include discoveries, inventions and works, as well as the birth and death dates of important thinkers.

MEMORY WORK

☐ Begin memorizing *List F: Major Bones of the Human Body*.

LESSON 31: NERVOUS SYSTEM

READING & RESEARCH

☐ Read KSE "The Brain and Nervous System", pp. 108-109.

☐ Read USE "The Nervous System", pp. 364-365.

☐ Read USE "The Brain", pp. 366-367.

DEFINITIONS

☐ In DB, read the definitions in "The Central Nervous System" p. 302-303.

☐ Define the following terms, noting any Latin or Greek word parts, and place in the Definitions section of your Science Notebook:

- central nervous system
- hormone
- brain
- spinal cord

LABS & ACTIVITIES

☐ Do Lab 31: Touchy Feely – BEK experiment 93, pp. 196-197.

☐ Write a complete Lab Report and file in the Labs section of the Science Notebook.

REPORT

☐ Write a 2-3 paragraph report on what you've learned and file in the Research Reports section of your Science Notebook.

SCIENCE HISTORY

☐ Record the birth and death dates of Ivan Pavlov in the Dates section of your Science Notebook and note his contribution.

☐ Research the work of Paul Broca and note any significant dates you encounter.

☐ Record any other significant dates you've encountered in the Dates section of your Science Notebook and on the timeline. Include discoveries, inventions and works, as well as the birth and death dates of important thinkers.

MEMORY WORK

☐ Continue memorizing *List F: Major Bones of the Human Body.*

HUMAN ANATOMY

LESSON 32: SKIN AND TOUCH

READING & RESEARCH

☐ Read KSE "Skin, Hair and Nails", pp. 100-101.

☐ Read KSE "Touch", p. 112.

☐ Read USE "Skin, nails and hair", pp. 368-369.

DEFINITIONS

☐ In DB, read the definitions in "The Skin", pp. 310-311.

☐ Define the following terms, noting any Latin or Greek word parts, and place in the Definitions section of your Science Notebook:

- *cutis*
- *integument*
- *dermis*
- *epidermis*

LABS & ACTIVITIES

☐ Do Lab 32: It's Cool – BEK experiment 90, pp. 190-191.

> ➤ Try this experiment with two thermometers. To more strongly see the effect of evaporation on cooling, do not use the alcohol soaked cotton on one of them.

☐ Write a complete Lab Report and file in the Labs section of the Science Notebook.

REPORT

☐ Write a 2-3 paragraph report on what you've learned and file in the Research Reports section of your Science Notebook.

SCIENCE HISTORY

☐ Research Sir Francis Galton and record the dates of his birth and death, along with the publication of his work *Finger Prints*.

☐ Record any other significant dates you've encountered in the Dates section of your Science Notebook and on the timeline. Include discoveries, inventions and works, as well as the birth and death dates of important thinkers.

MEMORY WORK

☐ Continue memorizing *List F: Major Bones of the Human Body*.

LESSON 33: TASTE AND SMELL

READING & RESEARCH

- ☐ Read KSE "Taste and Smell", p. 113.

- ☐ Read USE "The Nose and Tongue", pp. 374-375.

DEFINITIONS

- ☐ In DB, read the definitions in "Nerves and Nervous Pathways: Nose", p. 307.

- ☐ Define the following terms, noting any Latin or Greek word parts, and place in the Definitions section of your Science Notebook:

    o *nasal*

    o *oral*

    o *lingual*

    o *olfactory*

LABS & ACTIVITIES

- ☐ Do Lab 33: Totally Tasteful – BEK experiment 70, pp. 150-151.

    ➢ Two fingers are more comfortable than a clothespin.

    ➢ Try thoroughly mashing the samples so that the texture is not an immediate giveaway.

- ☐ Write a complete Lab Report and file in the Labs section of the Science Notebook.

REPORT

- ☐ Write a 2-3 paragraph report on what you've learned and file in the Research Reports section of your Science Notebook.

HUMAN ANATOMY

SCIENCE HISTORY

☐ Record any significant dates you've encountered in the Dates section of your Science Notebook and on the timeline. Include discoveries, inventions and works, as well as the birth and death dates of important thinkers.

MEMORY WORK

☐ Continue memorizing *List F: Major Bones of the Human Body.*

HUMAN ANATOMY

LESSON 34: EARS AND HEARING

READING & RESEARCH

- ☐ Read KSE "Ears, Hearing and Balance", pp. 116-117.

- ☐ Read USE "Ears", pp. 372-373.

DEFINITIONS

- ☐ In DB, read the definitions in "The Ears", pp. 314-315.

- ☐ Define the following terms, noting any Latin or Greek word parts, and place in the Definitions section of your Science Notebook:
    - *auditory*
    - *cochlea*
    - *malleus*
    - *incus*
    - *stapes*

LABS & ACTIVITIES

- ☐ Do Lab 34: Can You Hear Me Now? – BEK experiment 81, pp. 172-173.
    - ➢ Any twine or heavy string will do.
    - ➢ Try the activity with different sizes of utensils.

- ☐ Write a complete Lab Report and file in the Labs section of the Science Notebook.

REPORT

- ☐ Write a 2-3 paragraph report on what you've learned and file in the Research Reports section of your Science Notebook.

SCIENCE HISTORY

☐ Research Harvey Fletcher and his contribution.

☐ Record any other significant dates you've encountered in the Dates section of your Science Notebook and on the timeline. Include discoveries, inventions and works, as well as the birth and death dates of important thinkers.

MEMORY WORK

☐ Continue memorizing *List F: Major Bones of the Human Body.*

HUMAN ANATOMY

LESSON 35: EYES AND SIGHT

READING & RESEARCH

- ☐ Read KSE "Eyes and Seeing", pp. 114-115.
- ☐ Read USE "Eyes", pp. 370-371.

DEFINITIONS

- ☐ In DB, read the definitions in "The Eyes", pp. 312-313.
- ☐ Define the following terms, noting any Latin or Greek word parts, and place in the Definitions section of your Science Notebook:
  - *ocular*
  - *retina*

LABS & ACTIVITIES

- ☐ Do Lab 35: I Can See Clearly Now - BEK experiment 71, pp. 152-153
  - ➢ Any white paper without lines should work.
- ☐ Write a complete Lab Report and file in the Labs section of the Science Notebook.

REPORT

- ☐ Write a 2-3 paragraph report on what you've learned and file in the Research Reports section of your Science Notebook.

SCIENCE HISTORY

- ☐ Research who invented bifocal eyeglasses and note the name and date.

HUMAN ANATOMY

☐ Record any other significant dates you've encountered in the Dates section of your Science Notebook and on the timeline. Include discoveries, inventions and works, as well as the birth and death dates of important thinkers.

MEMORY WORK

☐ Recite List F: Major Bones of the Human Body for your family.

☐ File List F in the Memory Work section of your Science Notebook

LESSON 36: THE REAL WORLD WIDE WEB

READING & RESEARCH

☐ Read KSE "Biomes and Habitats", pp. 68-69.

☐ Read USE "Ecology", pp. 330-331.

☐ Read USE "Conservation", pp. 336-337.

☐ Read DSE Introduction to Ecology section, p. 369.

☐ Read DSE "The Biosphere", pp. 370-371.

☐ Read DSE "Conservation", p. 400.

DEFINITIONS

☐ In DB, read the definitions in "Living Things and Their Environment", p. 232.

☐ In DB, read the definitions in "Within an Ecosystem", p. 234.

☐ Define the following terms, noting any Latin or Greek word parts, and place in the Definitions section of your Science Notebook:

- ecology
- biosphere
- biome

LABS & ACTIVITIES

☐ Do Lab 36: Keepin' it Clean - BEK experiment 65, pp. 138-139.

> ➢ If you store your science supplies in a clear plastic tub, it can be temporarily emptied and used in lieu of the clear glass bowl for this experiment.

☐ Write a complete Lab Report and file in the Labs section of the Science Notebook.

REPORT

☐ Write a 2-3 paragraph report on what you've learned and file in the Research Reports section of your Science Notebook.

SCIENCE HISTORY

☐ Record the birth and death dates of Ernst Haeckel along with his contribution and its date.

☐ Record the birth date of James E. Lovelock and note his significance.

☐ Record any other significant dates you've encountered in the Dates section of your Science Notebook and on the timeline. Include discoveries, inventions and works, as well as the birth and death dates of important thinkers.

MEMORY WORK

☐ Challenge: Memorize *List G: Major Biomes of the Earth* and recite for your family in only one day!

☐ File *List G* in the Memory Work section of your Science Notebook.

LABS

LAB 01: MADNESS TO METHOD   Date:

BACKGROUND

Read "Plant Information" on the instruction sheet that came with your Botanical Discoveries kit.

OBJECTIVE

To apply the scientific method in testing which seeds will sprout first, mustard plant, sensitive plant, or cacti.

MATERIALS

- ☐ 6-8 mustard seeds
- ☐ 6-8 sensitive plant seeds
- ☐ 6-8 cacti seeds
- ☐ 1 ½ cups water
- ☐ 1½ cup germination mixture
- ☐ 3 germination cups
- ☐ 3 germination cup lids
- ☐ 3 plant stakes
- ☐ fine tip marker
- ☐ Lab Report forms
    - o Hypothesis
    - o Conclusion
    - o custom Results for this lab

© 2010 Classical Education Resources, LLC.   All rights reserved.

LAB 01: MADNESS TO METHOD

Date:

PROCEDURE

1. Complete the Hypothesis page of the Lab Report.

2. Fill each germination cup with ½ cup germination mixture. Break up or remove any large clumps.

3. Add 1/4 cup water to each cup and mix.

4. Place 6-8 mustard seeds in one container, 6-8 sensitive plant seeds in another and 6-8 cacti seeds in the third. Place them on top of the germination mixture about 3/8 of an inch apart, then press gently into the mixture with fingertips.

5. Carefully add more water on top of seeds being careful not to disturb them.

6. Label 3 planting markers and place in the appropriate cups.

7. Place lids on planting cups and place all 3 in a bright, sunny location.

8. Observe the cups each day for 10 days and look for any sprouting. Collect your data on the special Results form for this lab. Note the dates and the number of sprouts observed for each plant on each day. Circle the sprout number in the data cell for the first sprout you see and note this data in your results.

9. Complete the Lab Report by detailing your findings on the Conclusion form.

LABS

| LAB 01: MADNESS TO METHOD | Date: |

RESULTS

# Sprouts	Day									
	1	2	3	4	5	6	7	8	9	10
Sensitive										
Mustard										
Cacti										

NOTES

_____

_____

© 2010 Classical Education Resources, LLC.   All rights reserved.

LAB 02: CATEGORICALLY SPEAKING

Date:

BACKGROUND

Scientists classify living things based upon the organism's characteristics. They divide them into seven large groups, and then further divide them into subgroups based upon their differences. Each successive subgroup is more specific and contains fewer organisms. The last two subgroups, *genus* and *species*, can be used to identify a particular type of organism.

OBJECTIVE

To explore classification systems by grouping specimens according to their characteristics.

MATERIALS

- ☐ colored pencils – red, blue, yellow and green
- ☐ scissors
- ☐ copy of Handout 02-1, to color and cut
- ☐ Lab Report forms
    - o Conclusion
    - o custom Results for this lab (3 pages)

PROCEDURE

1. Color the specimen cards on the Lab Handout according to the column headings. Cut them out.

2. Lay the cards out on a table, floor or other large surface. What different features do you see in the specimens? Note your answer in question 1 on page 1 of the custom Results form.

3. Use the specimen cards to answer questions 2-6.

© 2010 Classical Education Resources, LLC. All rights reserved.

LAB 02: CATEGORICALLY SPEAKING

Date:

4. In order to design a classification system for the specimens you need to divide the cards into groups and subgroups according to any one of the distinguishing features you've observed about them. Choose one feature as the first level of grouping. Write the characteristic in the first column and first row of the Classification Possibilities grid on the Results form. For instance, if you choose to first divide them into two groups, large ones versus small ones, write "size" in the blank.

5. Next, choose another feature for the second level of grouping. Subdivide each of the big groups from step 4 into smaller groups according to that feature. Note your subdivision choice in the second column of the grid.

6. Finally, further subdivide these secondary groups into even smaller subgroups according to another feature. Write your third subdivision in column three. As you can see, the more we subdivide our specimens, the fewer we have in each group until we are able to zero in on a specific type of specimen.

7. Repeat steps 5 –7 to figure out five more classification possibilities and write each one on another row of the grid.

8. Does one method seem better than others? Answer question 7 on the second page of the Results form.

9. On the third page of the Results form, color the shapes in the first column to match your actual specimens. Lay out your cards on the floor to match the first column on the form.

10. Choose one of the classification systems you've created in the Classification Possibilities grid. Divide the cards according to the first grouping. Draw and color the specimen layout in the second column of the form. Label the top of the column to designate which feature you first divided by. Be sure to mark the dividing line between the groups. For example, if your first division gives you two groups, draw a line in the middle of the column to divide it into an upper and a lower section. If you need three groups, draw a line that divides the column into thirds. If you need four groups, mark the lines that divide the column into four groups.

11. Next, layout the cards to match your second classification feature. Label the top of the third column with the feature name. Draw the layout of your cards and draw the dividing line between your sub-groupings.

12. Label the last column with the feature you would finally sort by. Draw your objects in the final column. In this example we do not have any duplicates, so there will only be one object in each "group".

13. Can you think of another feature that some of the specimens have in common? Answer question 8 on your Results form.

14. Complete the Lab Report by detailing your findings on the Conclusion form.

LABS

LAB 02: CATEGORICALLY SPEAKING Handout 02-1

(1 COPY FOR EACH STUDENT, TO BE COLORED AND CUT OUT)

RED	BLUE	YELLOW	GREEN
○	○	○	○
∘	∘	∘	∘
□	□	□	□
▫	▫	▫	▫
△	△	△	△
▵	▵	▵	▵

© 2010 Classical Education Resources, LLC.   All rights reserved.

LABS

| LAB 02: CATEGORICALLY SPEAKING | Date: |

RESULTS

1. What different characteristics do you notice about the specimens?

2. How many specimens are large?

3. How many specimens are blue?

4. How many specimens are triangles?

5. How many specimens are both large and blue?

6. How many specimens are large blue triangles?

CLASSIFICATION POSSIBILITIES:

First by...	Then by...	Finally by...

LAB 02: CATEGORICALLY SPEAKING

Date:

RESULTS

7. Does one classification system seem more logical than the others? Why or why not?

_____
_____
_____
_____
_____

8. Look carefully at all the specimen cards. Can you discover other features that could be used to classify the organisms?

_____
_____
_____

LAB 02: CATEGORICALLY SPEAKING

Date:

RESULTS

ALL			

In the two middle columns, draw lines to divide your specimens into groups.

LAB 03: IT'S ALIVE! | Date:

BACKGROUND

Yeasts are fungi, living organisms with only a single cell. Like all living things, they have requirements to live, including food and water. When they digest their food, they give off carbon dioxide gas in a process called fermentation. This process is what makes bread dough rise.

The dried yeast purchased in a grocery store looks like powder or small grains. It does not look alive. This is because it has been dehydrated – the water has been removed.

OBJECTIVE

To discover whether dried yeast is still alive by providing food and water, and then looking for evidence of digestion.

MATERIALS

- [ ] dried yeast, 3 tsp
- [ ] sugar, 2 tsp
- [ ] 3 clear disposable cups, 16oz
- [ ] 2 tongue depressors/craft sticks/stirring spoons
- [ ] measuring cup, ¼ cup
- [ ] measuring spoon, 1-teaspoon
- [ ] warm water
- [ ] masking tape
- [ ] marker
- [ ] ruler, metric
- [ ] timer

| LAB 03: IT'S ALIVE! | Date: |

- ☐ Lab Report forms
    - o Hypothesis
    - o Conclusion
    - o custom Results for this lab

PROCEDURE

1. Complete the Hypothesis page of the Lab Report.
2. Place masking tape vertically up the side of each cup. Number the cups from 1 to 3 at the top of the masking tape strip near the cup rim.
3. Put 1 teaspoon of yeast in each cup.
4. Put 2 teaspoons of sugar into cup 3 only.
5. Put ¼ cup of warm water into cups 2 and 3 only.
6. Label the initial level of contents on each cup as 0.
7. Set each cup on a level surface, side by side.
8. Mark the content level on the tape every 5 minutes for 30 minutes.
9. After 30 minutes, measure in millimeters the height of each of the marks you made on each of the cups.
10. Record the data on the custom Results form for this lab.
11. Complete the Lab Report by detailing your findings on the Conclusion form.

LABS

| LAB 03: IT'S ALIVE! | Date: |

RESULTS

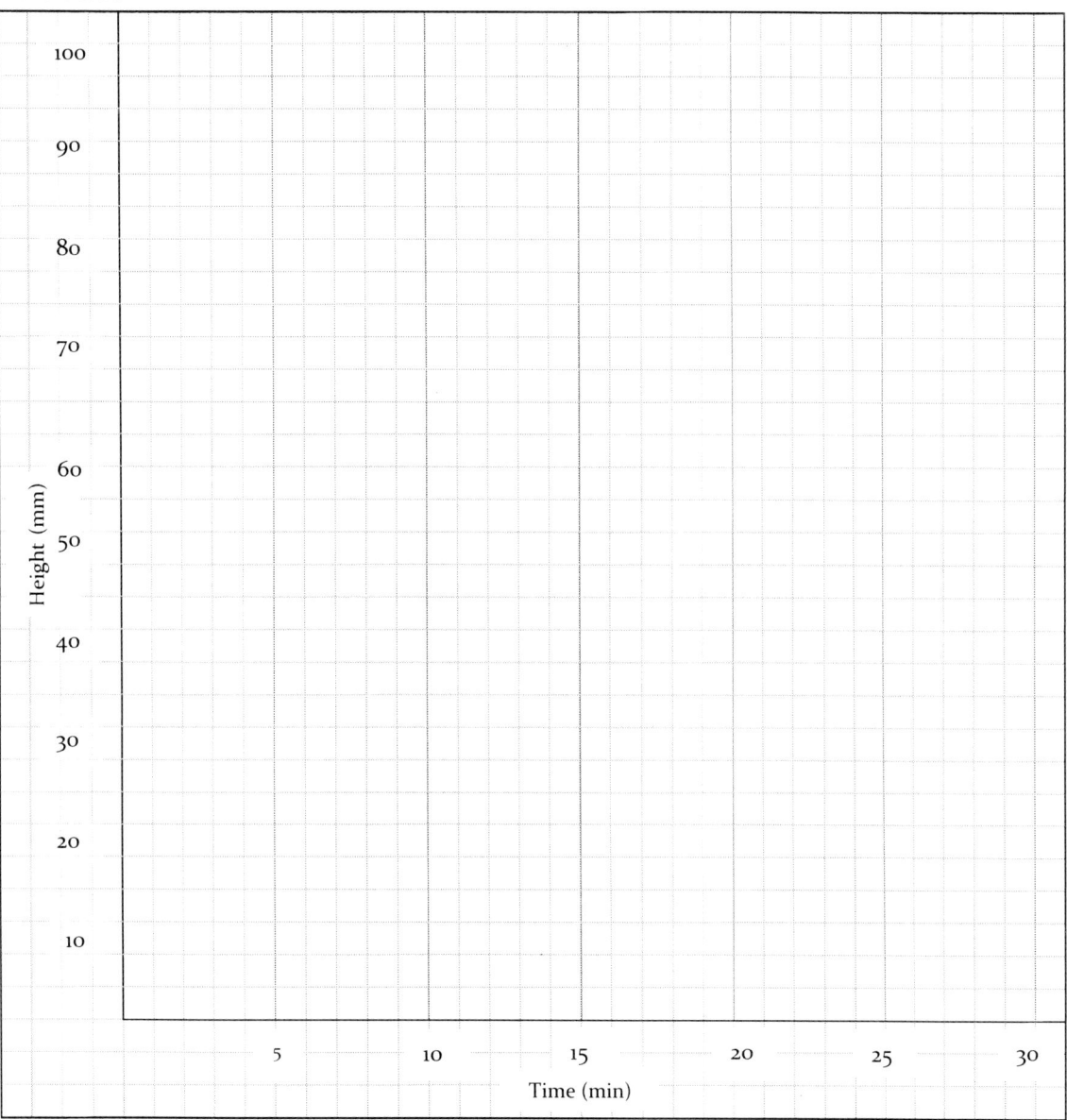

[Graph: Height (mm) vs Time (min); y-axis 0–100 mm, x-axis 0–30 min]

LEGEND: BLUE = YEAST ONLY;   GREEN = YEAST + WATER;   RED = YEAST + WATER + SUGAR

NOTES

© 2010 Classical Education Resources, LLC.   All rights reserved.

LABS

| **LAB 04: READY FOR MY CLOSEUP** | Date: |

BACKGROUND

The microscope will allow you to see the world to a degree of detail you might not have imagined.

OBJECTIVE

To learn how to use the microscope, prepare a wet mount slide and record your observations

MATERIALS

- ☐ microscope
- ☐ blank slide
- ☐ coverslip
- ☐ cork
- ☐ single edge razor blade
- ☐ tweezers
- ☐ pipette
- ☐ colored pencils
- ☐ Lab Report forms
    - o Results, Microscopy
    - o Conclusion

© 2010 Classical Education Resources, LLC.   All rights reserved.

LAB 04: READY FOR MY CLOSEUP

Date:

PROCEDURE

1. Have an adult cut a very thin section of the cork with a single-edge razor blade. Place a drop of water onto a clean slide. Gently place the slice of cork on the water drop using tweezers. Next, hold a cover slip beside the water drop at a 45 degree angle. Carefully lower the coverslip.

2. Set up the microscope as described the section titled *Using the Microscope* in this study guide. Remember to place it on a steady surface and at a comfortable viewing height.

3. Turn on the top light (incident light).

4. Rotate to the 4x objective until it clicks into place, giving 40x magnification.

5. Place the cork slide on the stage, under the stage clips, with the thinnest edge of the cork centered under the lens.

6. Carefully raise the stage to the highest level possible without touching the objective lens. Remember to look from the side while raising the stage.

7. Adjust the focus knob until the image is clear.

8. Switch to the 10x objective by rotating the nosepiece until it clicks into place, being careful not to touch the slide. Refocus for 100x magnification. View with both incident and transmitted lighting.

9. Next, change to the 40x objective to obtain 400x magnification. Experiment with both incident and transmitted lighting.

10. Finally, stain the cork specimen. Pull the stain using eosin y, following the directions in the *Using the Microscope* section of this study guide.

11. Using the Results, Microscopy form, make a sketch of the most interesting image.

12. Using the Conclusion form, write a brief description of your observations.

LABS

| **LAB 05: SAME DIFFERENCE** | Date: |

BACKGROUND

Plant and animal cells share most characteristics but have some very important differences.

OBJECTIVE

To document the similarities and differences between plant and animal cells.

MATERIALS

- ☐ microscope
- ☐ blank slides, 2
- ☐ coverslips, 2
- ☐ eosin Y stain
- ☐ methylene blue stain
- ☐ tweezers
- ☐ pipette
- ☐ petri dish or small disposable cup
- ☐ small onion
- ☐ colored pencils
- ☐ tongue depressor or craft stick
- ☐ copies of Handouts 05-1 and 05-2 to study

LAB 05: SAME DIFFERENCE Date:

☐ Lab Report forms

  o Results, Microscopy (2)

  o Conclusion

PROCEDURE

1. Have an adult cut a very small piece of onion. Using tweezers, peel the thin skin from the inside area of the onion piece. This is the concave side. The skin should be about as thin as clear food wrap.

2. Stain the onion skin with eosin y using the dip and rinse method detailed in the *Using the Microscope* section of this study guide.

3. Place a small drop of water onto a clean slide. Gently place the stained onion skin into the drop of water and add a coverslip as in the previous experiment.

4. Set up the microscope as detailed in the *Using the Microscope* section of this study guide. Turn on the bottom (transmitted) light and adjust the aperture to the widest opening. Rotate to the 4x objective.

5. Place the onion membrane slide on the stage. While looking from the side, carefully raise the stage to the highest level without actually touching the objective lens. Adjust the focus knob until you have a clear image. Try adjusting the aperture to lower light settings until you get the clearest image with sharpest contrast.

6. Move the slide around until you've centered the part of the specimen you want to examine further. Note the shape of the cells.

7. Rotate to the 10x objective, obtaining 100x magnification. Refocus. You may need to adjust the aperture for more light until you get the clearest image.

8. Repeat the steps for the 40x objective lens to get 400x magnification. You will probably require the maximum aperture opening for this magnification. Note the cell wall and the position of any visible nuclei. Do you see any chloroplasts? Can you make an educated guess as to why?

9. Using a Results, Microscopy form, make a sketch of what you see at 400x magnification.

10. Next, make a smear of cheek cells. To do this, use a clean craft stick or tongue depressor to gently scrape the inside of your cheek. You should scrape about 25 times to get a good sample. Spread the gathered cells in the center of a clean slide. Hold a coverslip perpendicular to the slide and wipe the edge to spread the cheek material. Let it dry a bit, and then add a drop of methylene blue stain. Gently place a coverslip by holding it perpendicular to the slide and lowering it gently onto the smear.

11. Reset the microscope by choosing the 4x objective, the widest aperture and the highest stage level that can be obtained without touching the objective lens.

12. What shape are these "animal" cells? Can you see a cell membrane? Move the slide around to center on area where some individual cells are visible. Switch to the 10x objective for 100x magnification. Can you make out details you could not see before? Where are the nuclei located?

13. Finally, switch to the 40x objective to obtain 400x magnification. Prepare a Lab Results, Microscopy form by sketching what you see using colored pencils.

14. Using the Conclusion form, write a brief description of your observations of the similarities and differences between plant and animal cells.

LABS

| LAB 05: SAME DIFFERENCE | Handout 05-1 |

Plant Cell

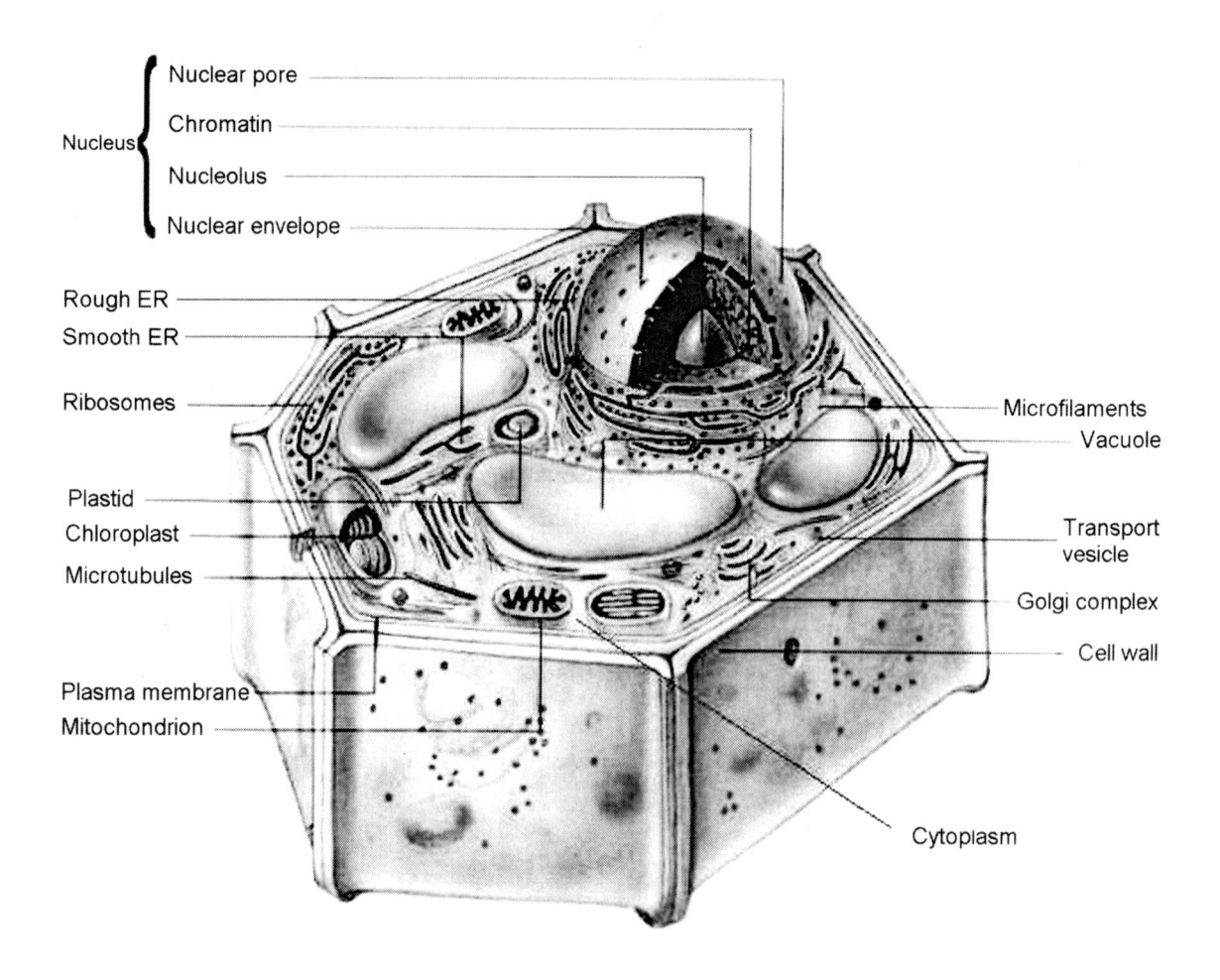

Nucleus
- Nuclear pore
- Chromatin
- Nucleolus
- Nuclear envelope

Rough ER
Smooth ER

Ribosomes

Plastid
Chloroplast
Microtubules

Plasma membrane
Mitochondrion

Microfilaments
Vacuole

Transport vesicle
Golgi complex
Cell wall

Cytoplasm

Plant cell diagram to study

© 2010 Classical Education Resources, LLC.   All rights reserved.

~ 138 ~

LAB 05: SAME DIFFERENCE — Handout 05-2

Animal Cell

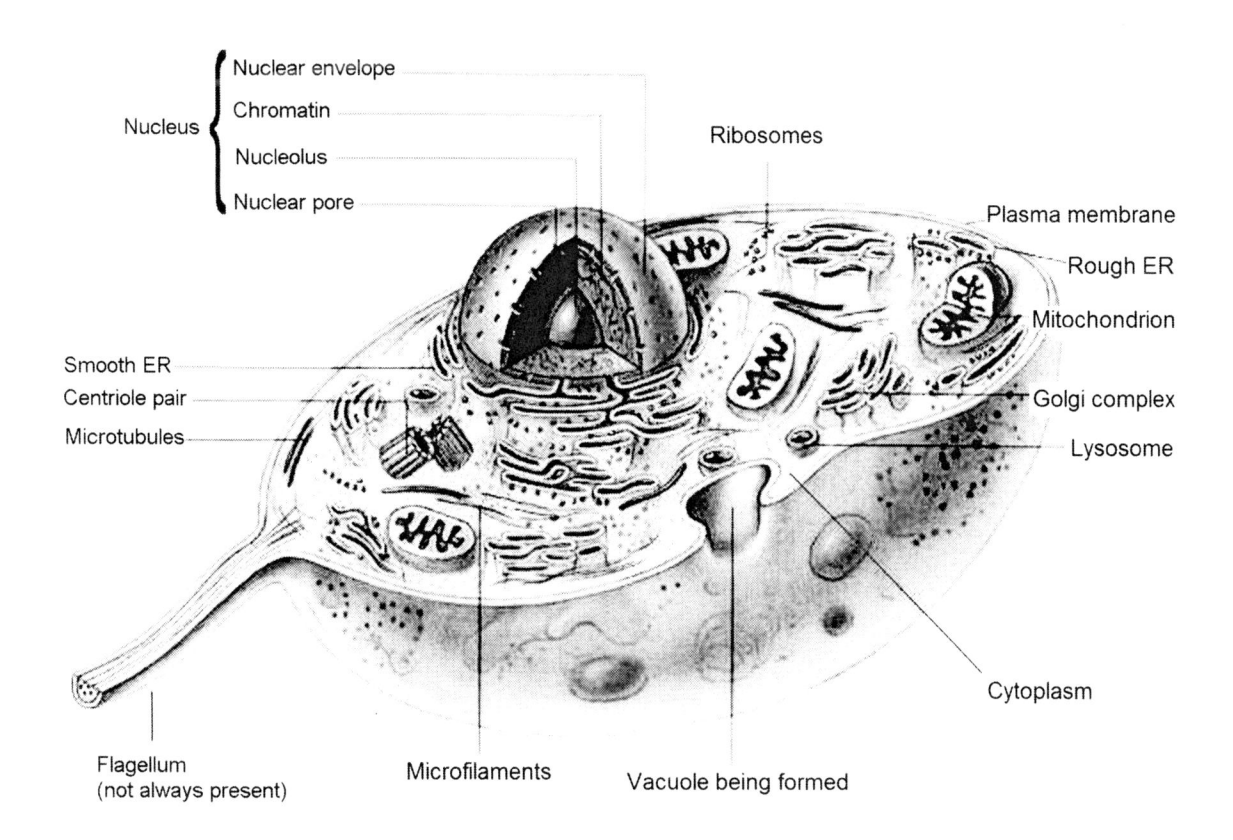

Nucleus:
- Nuclear envelope
- Chromatin
- Nucleolus
- Nuclear pore

Ribosomes
Plasma membrane
Rough ER
Mitochondrion
Smooth ER
Centriole pair
Microtubules
Golgi complex
Lysosome
Cytoplasm
Flagellum (not always present)
Microfilaments
Vacuole being formed

Animal cell diagram to study

LABS

| LAB 06: HYDRA HIJINX AND FUTURE FROGS | Date: |

BACKGROUND

Most living things create offspring through sexual reproduction or asexual reproduction.

OBJECTIVE

To observe asexual reproduction via budding in the hydra and the male and female gametes of sexual reproduction in frogs.

MATERIALS

- ☐ microscope
- ☐ prepared slides, hydra budding, frog sperm, frog ovary
- ☐ colored pencils
- ☐ Lab Report forms
    - o Results, Microscopy (3)
    - o Conclusion

PROCEDURE

1. Set up the microscope as detailed in the *Using the Microscope* section of this study guide. Turn on the bottom (transmitted) light and adjust the aperture to the widest opening. Rotate to the 4x objective.

2. Place the prepared hydra slide onto the stage. Center the hydra in the view. Raise the stage to the highest possible setting without hitting the objective lens and focus for the clearest view.

LAB 06: HYDRA HIJINX AND FUTURE FROGS — Date:

3. You should be able to see all or almost all of the animal. One end is the base that the hydra attaches to underwater plants and animals. At the other end are tentacles. The hydra uses the tentacles to capture and pull food into its mouth. Can you guess where the mouth is?

4. You should be able to see a lump projecting off the side of the hydra. This is the bud which will become an individual hydra. Budding is an example of asexual reproduction. Sometimes the bud will even have tentacles of its own, and sometimes there will be more than one bud at a time.

5. Using a Results, Microscopy form, draw and color what you see at the 40x magnification level you obtained with the 4x objective lens.

6. Next we will observe some organs of sexual reproduction. First, place the frog sperm slide on the stage of your microscope. Ensure that you are using the 4x objective, the bottom lighting and the widest aperture. Make sure the specimen is centered in the viewing area, and then raise the stage to the highest level available without touching the objective. Now look into the view and gradually bring the sperm smear into focus. You will see what looks like thin threads. The wider part is the head that contains the genetic material. The long, thin part is the tail that is used for swimming.

7. Observe the sperm at 100x and 400x magnification. Make a colored sketch of your observation of frog sperm at 400x magnification using a Results, Microscopy form.

8. Next we will view some organs of female sexual reproduction. Place the prepared slide of frog ovaries onto the microscope stage. Remember to reset the microscope to the lowest (4x) objective, highest aperture and highest stage setting that does not touch the objective lens.

9. Slowly bring the slide into focus. You should see small, immature follicles near the edges of the tissue. Toward the center are larger, mature follicles that contain darker ova in the center that are about to be released. You will also see follicles that have already released their eggs, leaving behind corpus lutea.

10. Next, observe the ovary at 100x and 400x magnification to see further details. Return to 40x magnification and make a colored drawing on a Results, Microscopy form. Label a follicle and an ovum.

11. Using the Conclusion form, write a brief description of your observations.

LABS

| LAB 07: SPLIT DECISION | Date: |

BACKGROUND

Cells grow and reproduce using two different kinds of splitting. Mitosis is used for growth and repair. Meiosis is used to produce gametes, the cells needed for reproduction. To ensure that the new organism receives only half of its genetic material from each parent, gametes must have only half the chromosomes of the parent cells.

OBJECTIVE

To document the similarities and differences between mitosis and meiosis.

MATERIALS

- ☐ microscope
- ☐ prepared slides, allium (onion) root tip showing mitosis, lilium undergoing meiosis
- ☐ colored pencils
- ☐ Lab Report forms
    - ○ Results, Microscopy (2)
    - ○ Conclusion

PROCEDURE

1. Set up the microscope as detailed in the *Using the Microscope* section of this study guide. Turn on the bottom (transmitted) light and adjust the aperture to the widest opening. Rotate to the 4x objective.

2. Place the prepared allium slide onto the stage. Find an interesting area of the specimen and center it in the view. Raise the stage to the highest possible setting without hitting the objective lens and focus for the clearest view.

© 2010 Classical Education Resources, LLC.   All rights reserved.

LAB 07: SPLIT DECISION

Date:

3. Adjust the aperture to lower light settings until you get the clearest image with sharpest contrast.

4. Choose the 10x objective lens by rotating to it until you hear the click. Refocus for 100x magnification. Adjust the aperture for more light (required when using higher magnification) until you get the clearest image. You should begin to distinguish some differences in the cell nuclei.

5. Repeat the steps for the 40x objective lens to get 400x magnification. At this magnification, you should be able to find nuclei in various stages of mitosis. Look for metaphase (genetic material lined up in the middle) and anaphase (chromosomes have migrated to opposite sides of the cell).

6. Try to find an area of the slide showing all four stages of mitosis. Using a Results, Microscopy form, draw and color what you see, labeling each stage of mitosis.

7. Reset the microscope to the lowest objective, widest aperture and highest stage level.

8. Look at the cutaway pictures of an anther on p. 257 and 258 of DB and p. 271 of USE.

9. Place the prepared lilium slide on the stage. Notice that the shape of the slide specimen is a cutaway version of the pictures you viewed. Draw and color what you see at 40x magnification on a Results, Microscopy form.

10. Center one of the pollen sacs. View at 100x and 400x magnification. Pollen is the male gamete for flowers. In order to produce pollen, some cells must undergo meiosis. What do you see in the center of the pollen sac? Do you see evidence of meiosis? Look for a cell that has doubled and one that has become four (tetrad).

11. Using a Lab Report Conclusion form, summarize your observations of cell mitosis and cell meiosis.

LAB 08: A NEW PAIR OF GENES   Date:

BACKGROUND

You have read that each individual has two genes for a trait, one from each parent.

You also learned that some traits are dominant and some are recessive. If the individual has one dominant gene and one recessive gene the dominant gene will control the expression and the individual will show the dominant trait.

The genes for a trait can be represented by a letter. If the gene is dominant the letter is capitalized and written first. If the trait is recessive the same letter is used but it is written in lowercase.

For example, curly hair is dominant and straight hair is recessive. We can represent the genes by:

C = curly (dominant) c = straight (recessive)

Since each person has two genes, one from each parent, their genotype can be either

CC = two curly (dominant) = curly hair

cc = two straight (recessive) = straight hair

Cc = one of each (hybrid) = dominant prevails = curly hair

Since there are three possible genotypes for each person (CC, cc or Cc) and there are two parents, there are six possible parent combinations:

CC + CC	cc + cc	CC + cc
Cc + CC	Cc + cc	Cc + Cc

LAB 08: A NEW PAIR OF GENES   Date:

If we know the genotypes of the parents we can find the possible genotypes and their percentage of probability in the offspring by constructing a *Punnett square*. We do this by drawing a 2 x 2 grid and writing the letters of one parent's genes across the top as column headings and the other parent's along the left side as row headings. We then fill the grid with the gene symbols that intersect in that grid. For example, if both parents are of type Cc our grid would look like this:

Mother = Cc

	C	c
C	CC	Cc
c	Cc	cc

Father = Cc

Recall that if the dominant gene is present it will control the expression of the trait. It "wins" the square. Three of the gene pairs above have a capital C and therefore would result in curly hair. One square has two recessive genes signified by two lowercase letters. This offspring would have straight hair. So there's a 3 out of 4 (75%) chance for curly hair and a 1 in 4 (25%) chance for straight hair.

OBJECTIVE

To learn how to construct Punnett squares to predict the possible traits of offspring.

MATERIALS

- ☐ Lab Report forms
    - o Conclusion
    - o custom Results for this lab

| LAB 08: A NEW PAIR OF GENES | Date: |

PROCEDURE

1. Answer the questions on the custom Results form.

2. Complete the Lab Report by detailing your findings on the Conclusion form.

LAB 08: A NEW PAIR OF GENES

Date:

RESULTS

1. What type of hair do the example parents above have? Remember their genotypes are both Cc. _____

2. What if one parent has both genes Cc (hybrid) and the other has two dominant genes CC? Fill in the Punnett Square and determine the probability of curly hair in the children.

    Mother = Cc

	C	c
C		
C		

    Father = CC

    % chance of curly hair (C is present in square) _____

    % chance of straight hair (no C is present in square) _____

    What are the possible genotypes for these children? _____

3. What parental genotypes would guarantee only straight-haired children?

4. Father _____   Mother _____

5. Now let's have some fun. Edward and Bella have 400 (!) children. Using the following traits and genotypes, figure *approximately* how many children would express each trait.

Trait	Dominant	Recessive	Edward	Bella
Eye color	B=brown	b=blue	BB	bb
Hair color	D=dark	d=light	dd	Dd
Hair texture	C=curly	c=straight	Cc	Cc

LAB 08: A NEW PAIR OF GENES

Date:

RESULTS

Eye color:
Edward = BB
Bella = bb

```
     b   b
  ┌───┬───┐
B │   │   │
  ├───┼───┤
B │   │   │
  └───┴───┘
```

Hair color:
Edward = dd
Bella = Dd

```
     D   d
  ┌───┬───┐
d │   │   │
  ├───┼───┤
d │   │   │
  └───┴───┘
```

Hair type:
Edward = Cc
Bella = Cc

```
     C   c
  ┌───┬───┐
C │   │   │
  ├───┼───┤
c │   │   │
  └───┴───┘
```

6. Approximately how many children will have brown eyes? _____
7. Approximately how many children will have blue eyes? _____
8. Approximately how many children will have dark hair? _____
9. Approximately how many children will have light hair? _____
10. Approximately how many children will have curly hair? _____
11. Approximately how many children will have straight hair? _____
12. Their last (!) child has the following genotypes:

    Bb   dd   CC

13. What does she look like? _____

   For eyes, brown is dominant and blue is recessive. For hair color, dark is dominate and blond is recessive.

14. Can you guess your genotypes for eye and hair color? _____

15. Can you guess the possibilities for your parents' genotypes?

    Father's eye color _____   Father's hair color _____

    Mother's eye color _____   Mother's hair color _____

LABS

| **LAB 10: FUNGUS AMONG US** | Date: |

BACKGROUND

Mushrooms belong to a group of organisms known as fungi which are neither plants nor animals. They reproduce by a process known as *sporulation*.

OBJECTIVE

To determine whether mushrooms require some stimulation, such as wind or animal touch, to release their spores.

MATERIALS

- ☐ whole mushroom from the store (not wild – some are poisonous!)
- ☐ dark paper (use light colored for portabella mushrooms)
- ☐ disposable bowl
- ☐ art fixative or hairspray
- ☐ knife
- ☐ Lab Report forms
    - o Hypothesis
    - o Results, Observation
    - o Conclusion

PROCEDURE

1. Complete the Hypothesis form of the Lab Report.

2. Make sure the mushroom is dry. If it seems overly damp, turn it gill side up for 24 hours so that it can dry out a bit.

LAB 10: FUNGUS AMONG US — Date:

3. Cut a ½ inch hole in the bottom of the container to allow moisture to escape.

4. Cut the stem off the mushroom, even with the bottom of the cap. If the cap edges fold under a great deal, gently trim around the rim so that the gills are exposed.

5. Lay paper on smooth, stable surface, and place mushroom in center with gill side down. Turn bowl upside down and carefully place over mushroom. Leave undisturbed for 2 days.

6. After two days, carefully remove the bowl and lift the mushroom cap straight up.

7. Observe the design left by the mushroom cap.

8. Lightly spray fixative or hairspray on the design, following the directions on the package.

9. Cut out and paste the dried spore print onto a Results, Observation form.

10. Complete the Lab Report by detailing your findings on the Conclusion form.

LABS

| LAB 14: WHAT COT? | Date: |

BACKGROUND

Monocots have their vascular bundles arranged irregularly, while dicots have them arranged in an orderly fashion, usually in a ring.

OBJECTIVE

To compare monocot and dicot plants by examining cross sections of the vascular systems of their respective buds.

MATERIALS

- ☐ microscope
- ☐ prepared slide, 2 flower types
- ☐ colored pencils
- ☐ Lab Report forms
    - o Results, Microscopy (2)
    - o Conclusion

PROCEDURE

1. Set up the microscope as detailed in the *Using the Microscope* section of this study guide. Turn on the bottom (transmitted) light and adjust the aperture to the widest opening. Rotate to the 4x objective.

2. Place the prepared flower slide onto the stage. Your slide contains both monocot and dicot buds in cross section.

3. Find the monocot bud. How do you know it is a monocot? How many vascular bundles do you see? Using a Results, Microscopy form, draw and color what you see at 40x magnification.

© 2010 Classical Education Resources, LLC.   All rights reserved.

4. Now switch to the dicot specimen. How can you distinguish the dicot bud from the monocot bud? How many petals does your bud have? How many sepals? Can you see the xylem and phloem? Draw and color what you see at 40x on another Results, Microscopy form. Label the phloem, xylem and cambrium.

5. On the Conclusion form, detail the differences you noted between the moncot and dicot buds.

LAB 20: FOOTLOOSE

Date:

BACKGROUND

Insect legs have the same basic parts. However, various species have developed highly specialized leg structures that are suitable for their unique lifestyles and requirements.

Basic Structure of Insect Legs

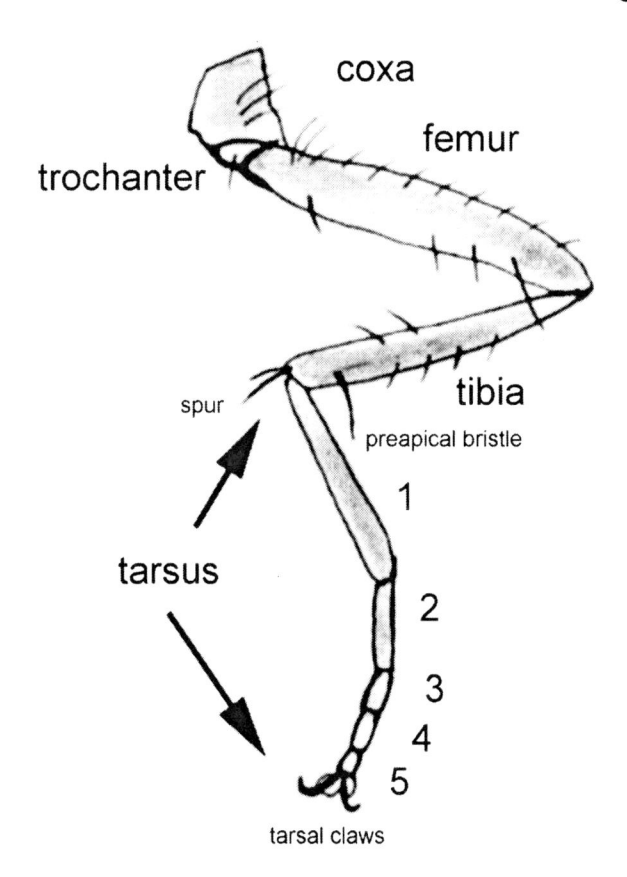

- coxa
- femur
- trochanter
- spur
- tibia
- preapical bristle
- tarsus
  - 1
  - 2
  - 3
  - 4
  - 5
- tarsal claws

Although the number of tarsal segments can vary, and tarsal claws are not always present, the basic structure is the same.

Some versions of leg specialization are: clinging, digging, swimming, running, jumping and grasping. The variations are detailed below.

Insect Leg Specialization

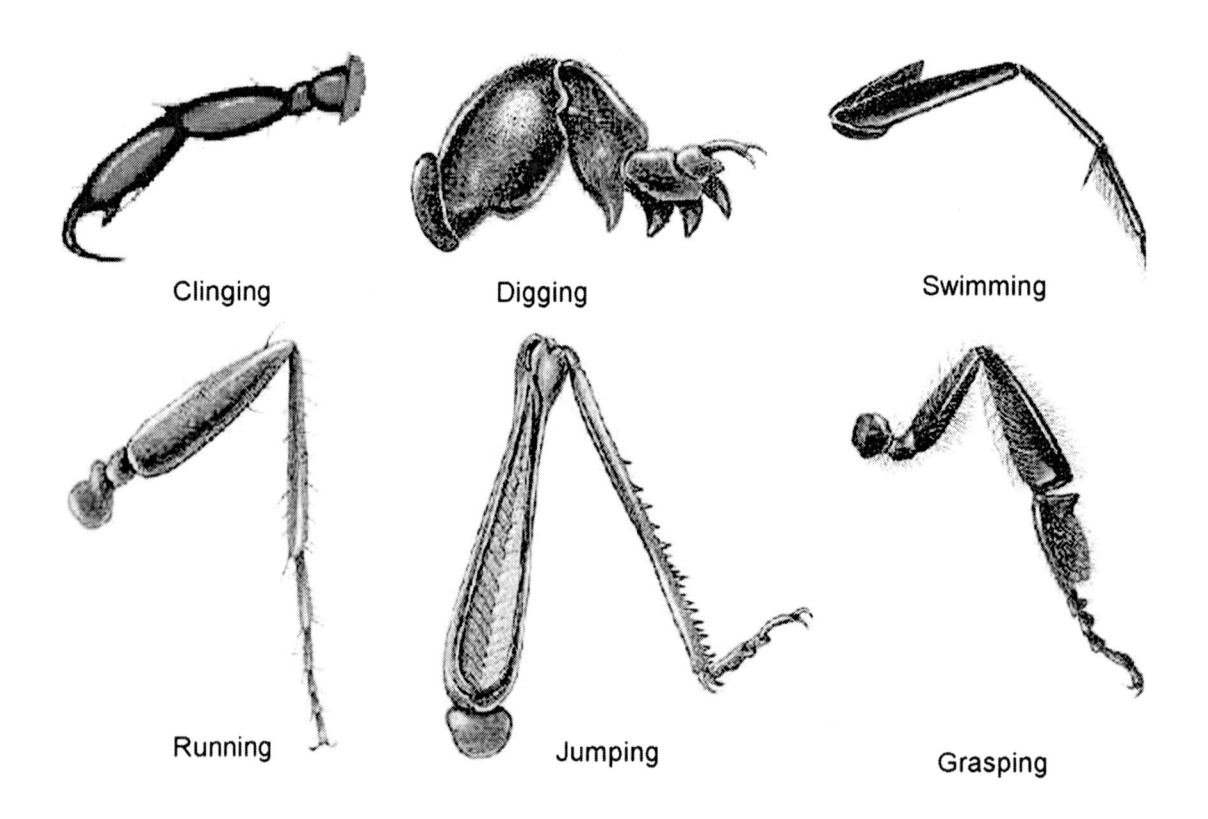

Clinging Digging Swimming

Running Jumping Grasping

clinging leg   claw shape and a barb opposite the claw to help with hanging on to hair (louse)

digging leg   trowel-like projections, short and bulky to aid with excavation (mole cricket)

swimming leg   slim with long hairs for buoyancy and movement through water (water strider)

running leg   long and lean for rapid movement (cockroach)

jumping leg   large, powerful thigh to give spring (grasshopper)

grasping leg   claws at the end and multiple segments for holding (honeybee)

LABS

OBJECTIVE

To observe four different types of insect leg anatomy

MATERIALS

- ☐ microscope
- ☐ prepared slide, 4 insect legs
- ☐ colored pencils
- ☐ Lab Report forms
    - ○ Results, Microscopy (4)
    - ○ Conclusion

PROCEDURE

1. Set up the microscope as detailed in the *Using the Microscope* section of this study guide. Turn on the bottom (transmitted) light and adjust the aperture to the widest opening. Rotate to the 4x objective.

2. Place the prepared insect leg slide onto the stage. Your slide contains four types of insect legs.

3. Look at each of the four specimens in turn. Using a separate Results, Microscopy form for each specimen, draw and color what you see at 40x magnification.

4. After drawing the insect legs you viewed, lay the drawings on a table side by side. Take note of any differences in the leg parts. Try to determine which type of leg each represents, and label it as such.

5. On the Conclusion form, detail the differences you observed between the four insect legs.

LABS

| LAB 21: FISHY BIRTHDAYS | Date: |

BACKGROUND

Have you ever heard an adult joke about being much younger than they actually are? Fish would have a hard time fibbing about their age. The scales of most bony fish show growth rings, much like trees. The growth of the scales is rapid during those times of year when food is more available, leading to a wider band. At those times of year when food is more scarce the scales grow slowly making the ring narrower.

OBJECTIVE

To estimate the age of a fish by counting growth rings on a scale.

MATERIALS

- ☐ microscope
- ☐ prepared slide, fish scales
- ☐ colored pencils
- ☐ Lab Report forms
    - o Results, Data
    - o Results, Microscopy
    - o Conclusion

PROCEDURE

1. Set up the microscope as detailed in the *Using the Microscope* section of this study guide. Turn on the bottom (transmitted) light and adjust the aperture to the widest opening. Rotate to the 4x objective.

2. Place the prepared fish scales slide onto the stage. Your slide contains three different kinds of scales. Two of them show growth rings. We are interested in examining the

cycloid scale. It is the one with relatively smooth edges, i.e.; no spiny projections on one side.

3. Note the light and dark rings. Find the area of the specimen with the widest light rings.

4. Now count the wide, light rings. Do this 3 times and average your result using a Results, Data form to hold your calculations. What is the approximate age of your fish? What do you think you would observe in an aquarium raised fish that experienced consistent temperatures and a constant food supply?

5. Draw the cycloid scale on a Results, Microscopy form using colored pencils.

6. Write what you learned on a Conclusion form. Approximately how old was your fish?

LAB 24: WHOOO EATS WHAT

Date:

BACKGROUND

Owls belong to the group of predatory birds known as raptors. They eat primarily small mammals, but when supply is short they may eat insects, birds and even reptiles.

OBJECTIVE

To learn about the diet of owls.

MATERIALS

- ☐ Owl Pellet Kit, includes
    - o owl pellet
    - o tweezers
    - o probe
    - o guide and worksheet
    - o bone sorting chart
    - o skull sorting key
- ☐ disposable gloves
- ☐ paper plates, 3
- ☐ Lab Report forms
    - o Hypothesis
    - o Conclusion
    - o custom Results for this lab will be a copy of the guide that came with the kit

LABS

PROCEDURE

1. Read both sides of the Owl Pellet Guide and Worksheet before beginning the dissection. This sheet will also serve as the Results page of your Lab Report.

2. Prepare your written hypothesis on the Hypothesis form. What do you think you will find in the pellet?

3. Perform the dissection following the instructions in the guide. Be gentle with the pellet as some of the bones are fragile. Carefully follow all safety guidelines.

4. Complete the Conclusion page of your Lab Report. What had your owl eaten?

LAB 25: HAIRY SITUATION

Date:

BACKGROUND

Mammals are the only class of vertebrates that have hair on their bodies. Hair has a three-part structure, consisting of a cuticle on the outside, a cortex which is the main body of the hair, and a medulla at the core.

Hair Structure

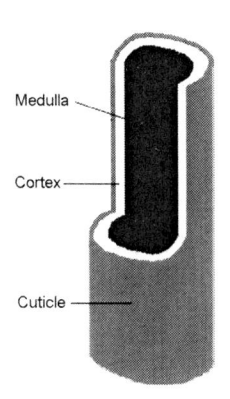

The medulla's features can help to identify the mammal that is the source of the hair. The medulla types indicate the spacing of the medulla cells in the hair.

Medulla Types

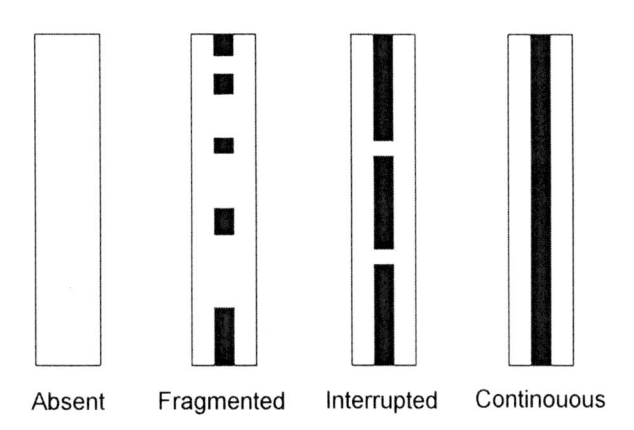

Absent   Fragmented   Interrupted   Continouous

Medullae can also exhibit certain patterns in their structure.

Medulla Patterns

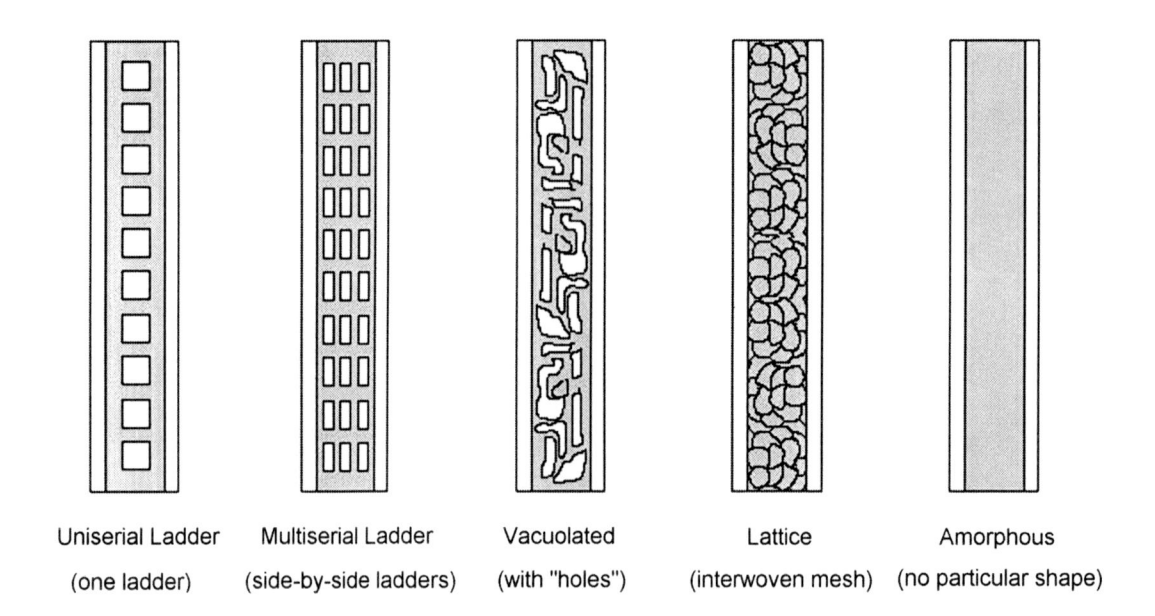

Uniserial Ladder	Multiserial Ladder	Vacuolated	Lattice	Amorphous
(one ladder)	(side-by-side ladders)	(with "holes")	(interwoven mesh)	(no particular shape)

The most useful characteristic for identification is the medullary index, which is a ratio of the width of the medulla to the width of the hair as a whole. In humans the medulla is usually less than 1/3 the width of the hair, if it is present at all. In animals it is usually greater than 1/3 the width, very often being greater than ½ the width.

Use the chart below for reference as you examine your specimens with the microscope.

	Medulla Type	**Medulla Pattern**	**Medullary Index**
Human	usually absent or fragmented; can be any type	usually absent or amorphous	usually 0 – 1/3
Cat	usually continuous	usually uniserial ladder, occasionally vacuolated	usually 1/3 – 9/10
Dog	usually continuous	usually amorphous to vacuolated	usually 1/3 – 9/10

OBJECTIVE

To observe and compare three different types of mammal hair.

MATERIALS

- ☐ microscope
- ☐ blank slides, 2-3
- ☐ coverslips, 2-3
- ☐ tweezers
- ☐ pipette
- ☐ human hair
- ☐ dog hair ('normal' hair from the back, not the downy underfur from the belly)
- ☐ cat hair ('normal' hair from the back, not the downy underfur from the belly)
- ☐ colored pencils
- ☐ Lab Report forms
    - o Results, Data
    - o Results, Microscopy (3)
    - o Conclusion

PROCEDURE

1. Prepare wet mount slides of the human, dog and cat hairs.

2. Prepare a Results, Data form by constructing a table with 4 rows and 4 columns. The top row will hold headings for Index, Type and Pattern. The leftmost column will contain headings for Human, Dog and Cat. Remember to leave the top left cell blank – start your column headings in column 2 and your row headings in row 2.

3. Set up the microscope as detailed in the *Using the Microscope* section of this study guide. Turn on the bottom (transmitted) light and adjust the aperture to the widest opening. Rotate to the 4x objective.

4. Place the human hair slide on the stage. Using the 4x objective, scan the hair from end to end. What type of medulla does it have, if any? Is it absent, fragmented, interrupted or continuous? Change to the 10x objective and find an area with medulla, if any. What is the relative width of the medulla? Is it less than 1/3 the width of the hair? Change to the 40x objective. Do the cells in the medulla exhibit a pattern? Draw and color what you see at 400x on a Results, Microscopy form.

5. Next, switch back to the 4x objective. Choose one of the animal hairs you have collected. Scan the entire length of the specimen. Can you detect what type of medulla this animal has? Move to the 10x objective. What is the medullary index for this hair? Switch to the 40x objective. Does the medulla have a pattern? Note your findings in the table you created on the Results, Data form. Draw and color what you see using the 40x objective on a Results, Microscopy form.

6. Finally, change back to the 4x objective. Place the other animal hair on the stage and scan it from end to end. What type of medulla does it have? Change to the 10x objective to determine the medullary index. Finally, observe the specimen using the 40x objective to clarify the medullary pattern. Document your observations in the table on the Results, Data form. Draw and color what you see at maximum magnification on a Results, Microscopy form.

7. Using the Conclusion form, detail the differences you observed in the various types of mammal hair. Did your hairs match the usual expectations?

LABS

| **LAB 27: WHAT'S YOUR TYPE?** | Date: |

BACKGROUND

Blood typing allows medical personnel to safely perform transfusions by ensuring that donors and recipients have the same or compatible blood types.

OBJECTIVE

To discover your own blood type.

MATERIALS

- ☐ Blood Typing Kit, includes
    - o blood typing card
    - o pipette
    - o alcohol prep pad
    - o sterile lancet
    - o mixing sticks, 4
- ☐ disposable gloves
- ☐ a helper
- ☐ Lab Report forms
    - o Results, Observation
    - o Conclusion

PROCEDURE

1. Carefully read both sides of the instruction guide that came with your kit before starting.

LABS

2. Prepare your written hypothesis. Do you know your blood type?

3. Have your helper put on the disposable gloves.

4. Perform the test following the instructions in the guide. *Carefully follow all safety guidelines!* The lancet is very sharp and blood may carry disease. Never touch someone else's blood with your bare hands, and never let others touch yours even if you think you are well

5. Using the Results, Observation form, draw what you see on the card. You may also document your results by taking a photograph. Carefully wrap and dispose of the card when you are finished.

6. Complete the Conclusion page of your Lab Report. What is your type?

LABS

| LAB 29: DON'T BE PUSHY | Date: |

BACKGROUND

Humans have three different types of muscle tissue. Each kind is specialized for the work it must do.

OBJECTIVE

To observe striated, smooth and cardiac muscle differences.

MATERIALS

- ☐ microscope
- ☐ prepared slide, 3 muscle types
- ☐ colored pencils
- ☐ Lab Report forms
    - o Results, Microscopy (3)
    - o Conclusion

PROCEDURE

1. Set up the microscope as detailed in the *Using the Microscope* section of this study guide. Turn on the bottom (transmitted) light and adjust the aperture to the widest opening. Rotate to the 4x objective.

2. Place the prepared muscle slide onto the stage. Your slide contains three types of muscle specimens. Read the rest of the steps outlined below and look at all three specimens before proceeding.

3. Find the striated muscle. The cells of striated muscle are very thin and long. They have multiple nuclei which are arranged near the edges of the cell. What distinguishes striated muscle cells from the smooth muscle cells is the presence of striations which

© 2010 Classical Education Resources, LLC.   All rights reserved.

~ 169 ~

look like stripes that run across the cell (the short way). These cells are specialized to allow voluntary movement of the muscles. Draw what you see using your colored pencils on one of the Results, Microscopy forms.

4. Next, look for the smooth muscle. Smooth muscle cells are somewhat elongated. Smooth muscle cells have only one nucleus per cell, which is generally of an oval shape. Most importantly, they exhibit a complete lack of striations. Draw the smooth muscle cells you viewed on another Results, Microscopy form.

5. Finally, find the cardiac muscle specimen. Cardiac muscle contains some of the features of smooth muscles and some of striated muscles. These cells have multiple nuclei and some striations across the cell. They will also display some dark bands that look like very wide striations. These are the intercalated discs that join the ends of the cardiac muscle cells together. Draw what you see, and label one cell's nuclei, striations and intercalated bands.

6. On the Conclusion form explain the differences you observed between the three types of muscle cells.

LAB 04: READY FOR MY CLOSEUP

Diagram 04-1

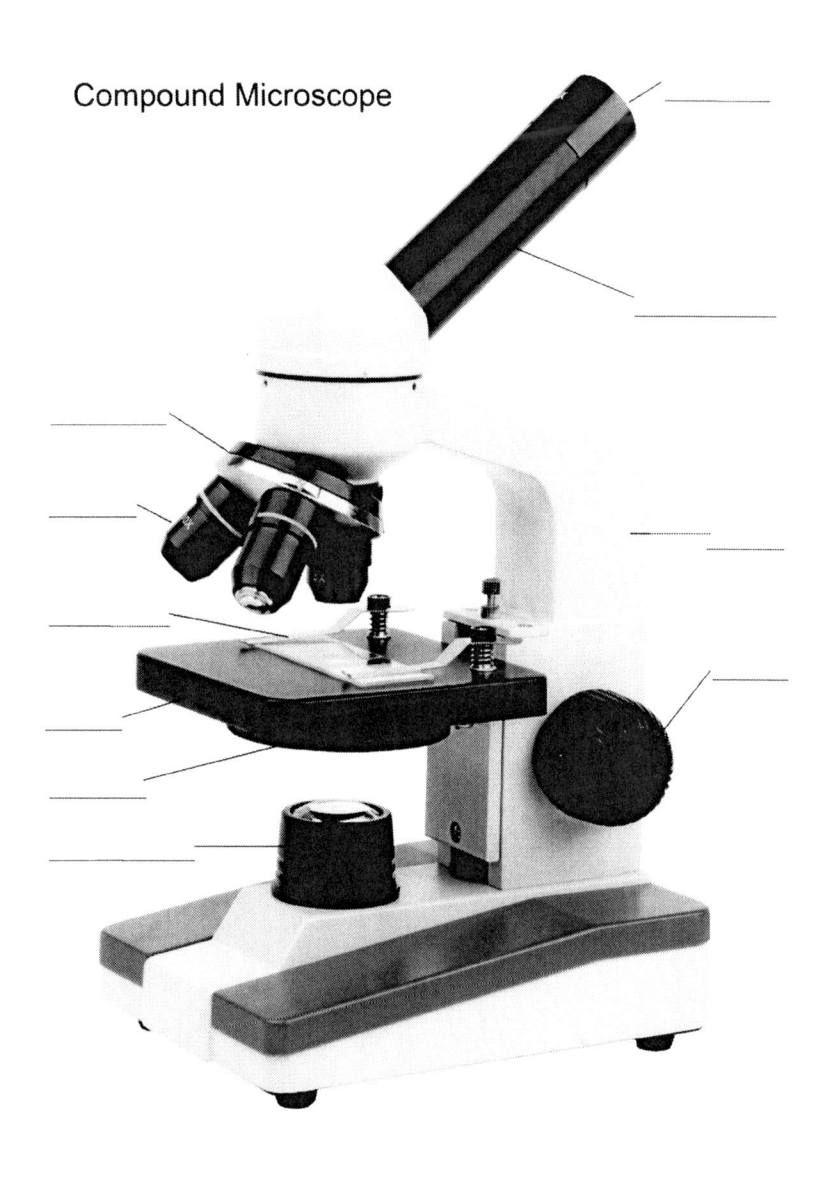

Compound Microscope

LAB 05: SAME DIFFERENCE Diagram 05-1

Plant Cell

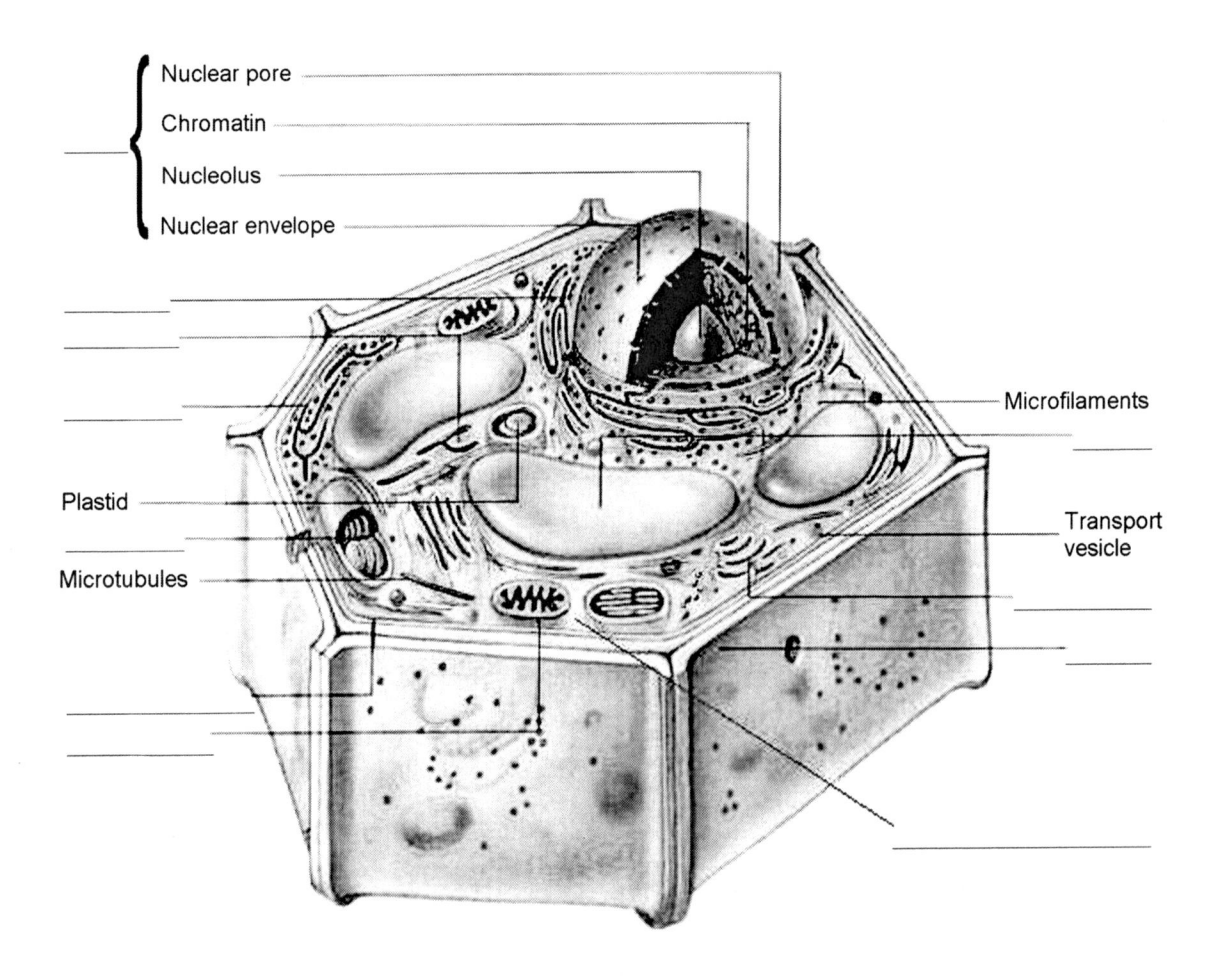

- Nuclear pore
- Chromatin
- Nucleolus
- Nuclear envelope
- Microfilaments
- Plastid
- Transport vesicle
- Microtubules

Label the horizontal lines above to denote these organelles:

- **A** Cell wall
- **B** Smooth ER
- **C** Cell membrane
- **D** Nucleus
- **E** Vacuole
- **F** Cytoplasm
- **G** Ribosome
- **H** Chloroplast
- **I** Rough ER
- **J** Mitochondrion
- **K** Golgi complex

*ER = Endoplasmic reticulum   Cell membrane is also known as plasma membrane

LAB DIAGRAMS

LAB 05: SAME DIFFERENCE | Diagram 05-2

Animal Cell

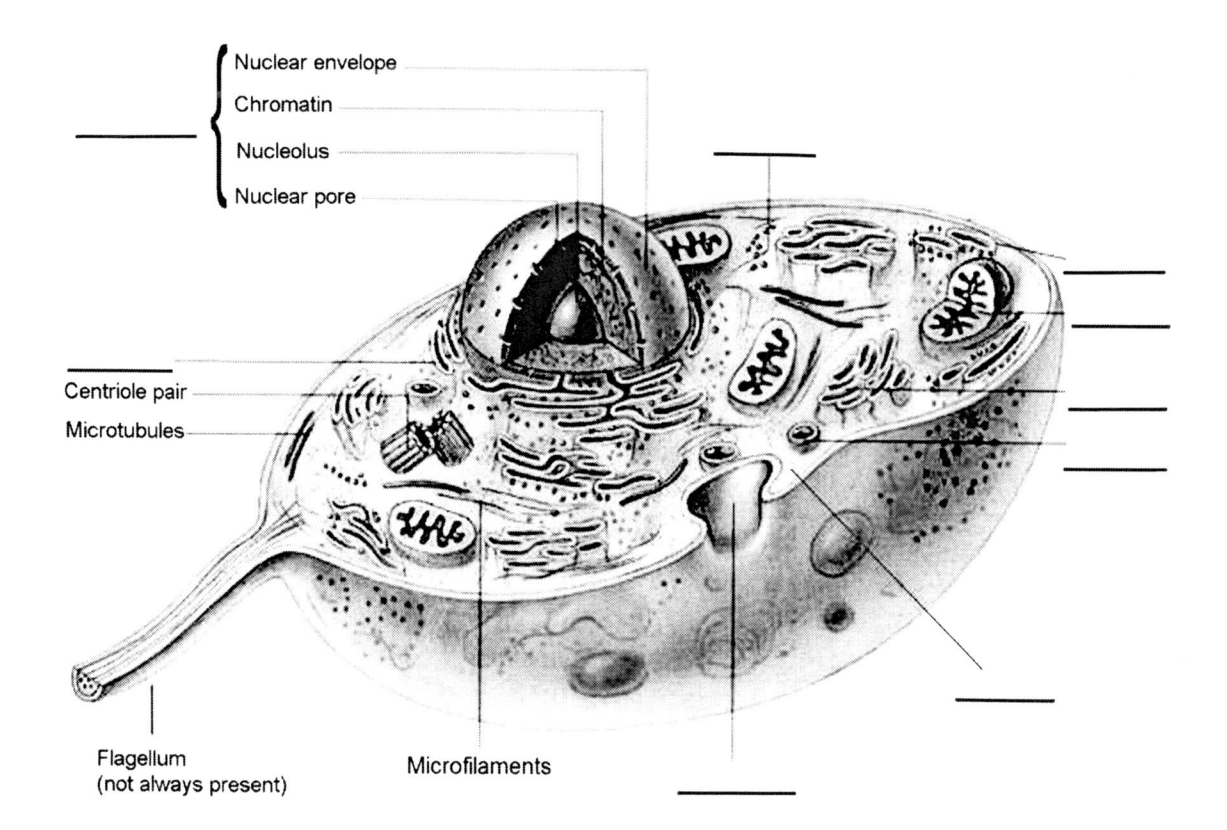

Nuclear envelope
Chromatin
Nucleolus
Nuclear pore

Centriole pair
Microtubules

Flagellum
(not always present)

Microfilaments

Label the horizontal lines above to denote these organelles:
- **A** Lysosome
- **B** Cell membrane
- **C** Smooth ER
- **D** Golgi complex
- **E** Mitochondrion
- **F** Ribosome
- **G** Vacuole
- **H** Rough ER
- **I** Nucleus
- **J** Cytoplasm

* ER = Endoplasmic reticulum   Cell membrane is also known as plasma membrane
* In this picture, vacuole is being formed on edge of cell

© 2010 Classical Education Resources, LLC.   All rights reserved.

Mitosis

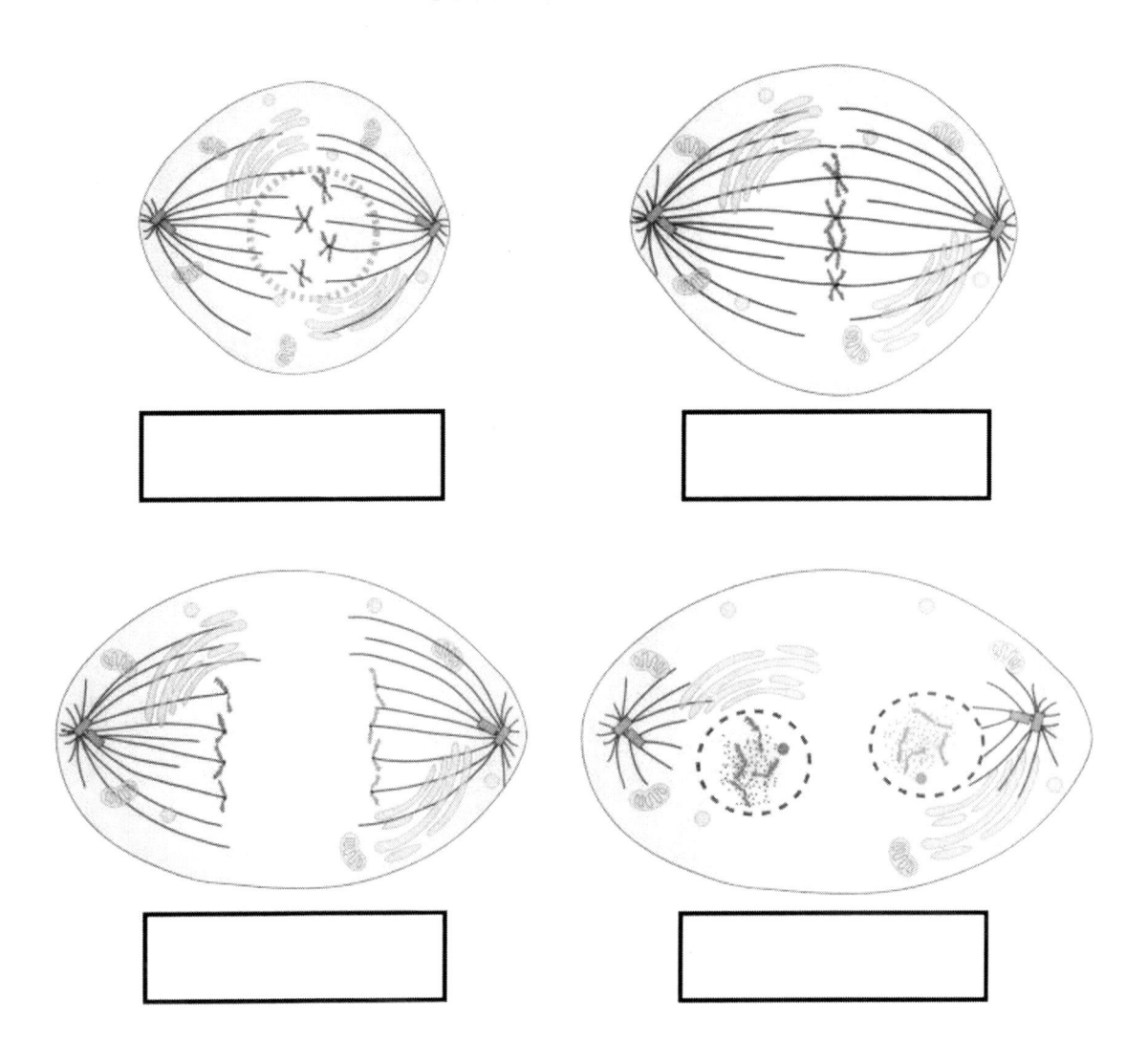

LAB DIAGRAMS

LAB 07: SPLIT DECISION — Diagram 07-2

Meiosis I

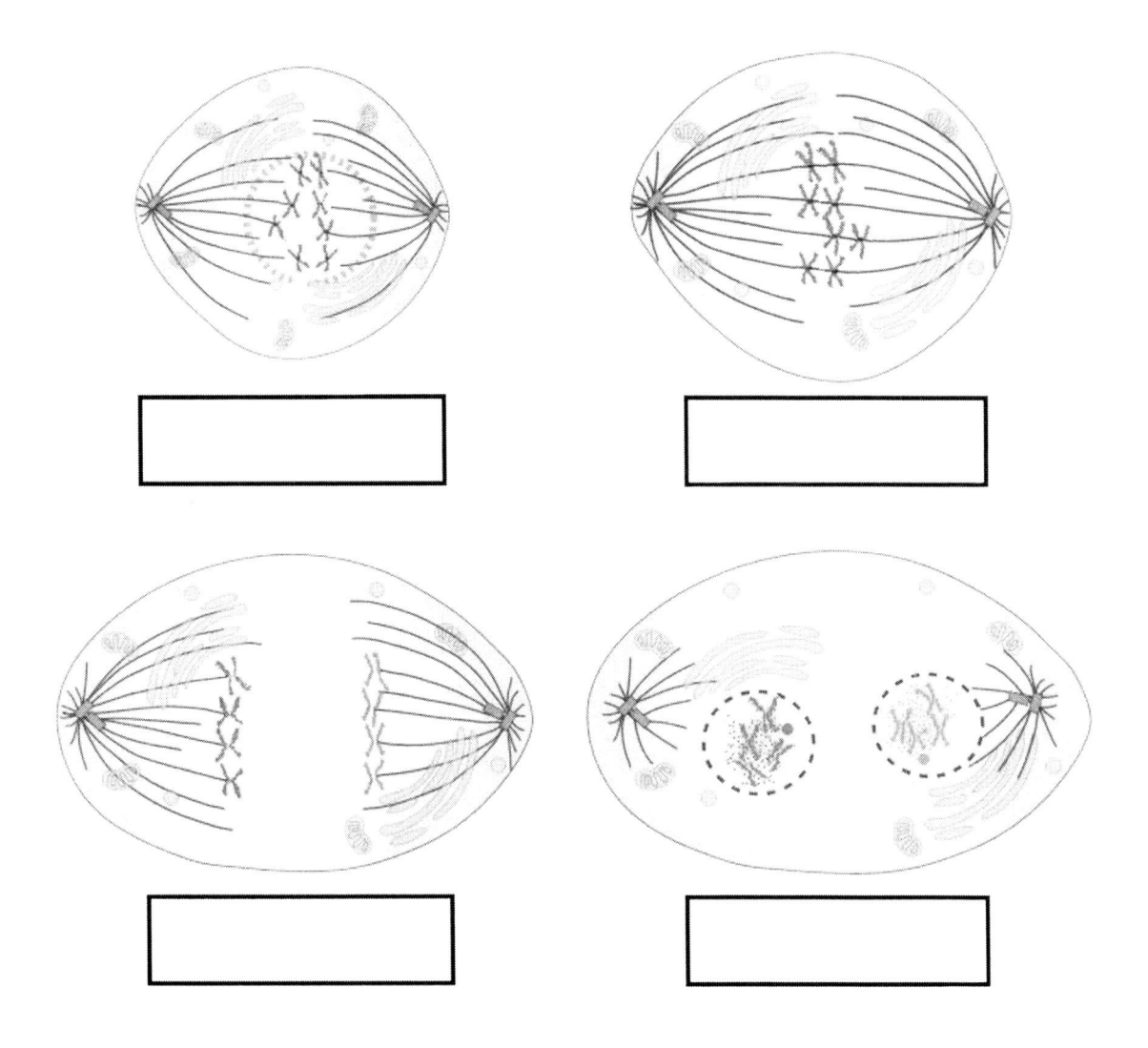

LAB DIAGRAMS

LAB 07: SPLIT DECISION   Diagram 07-3

Meiosis II

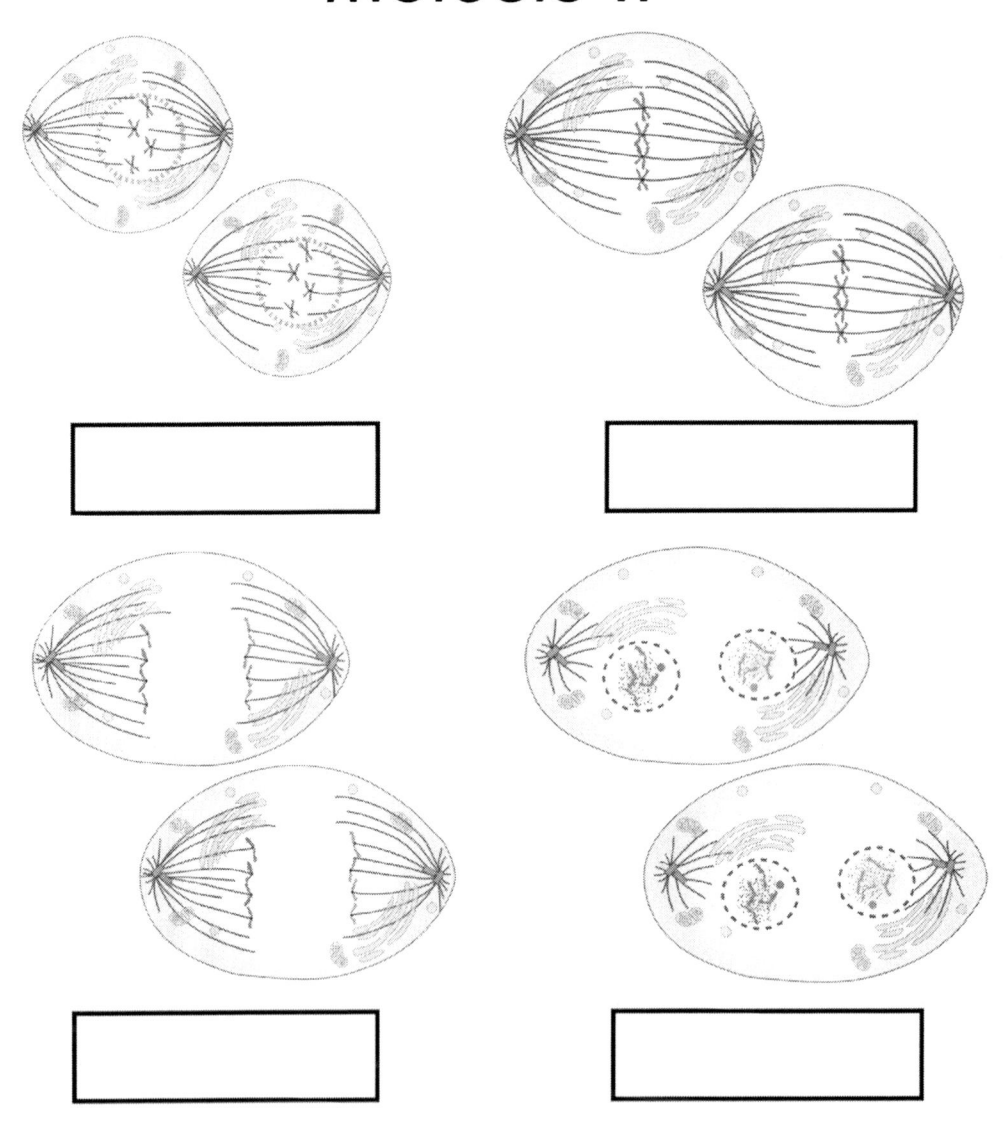

~ 176 ~

LAB DIAGRAMS

LAB 19: STUDY A SPIDER — Diagram 19-1

Tarantula

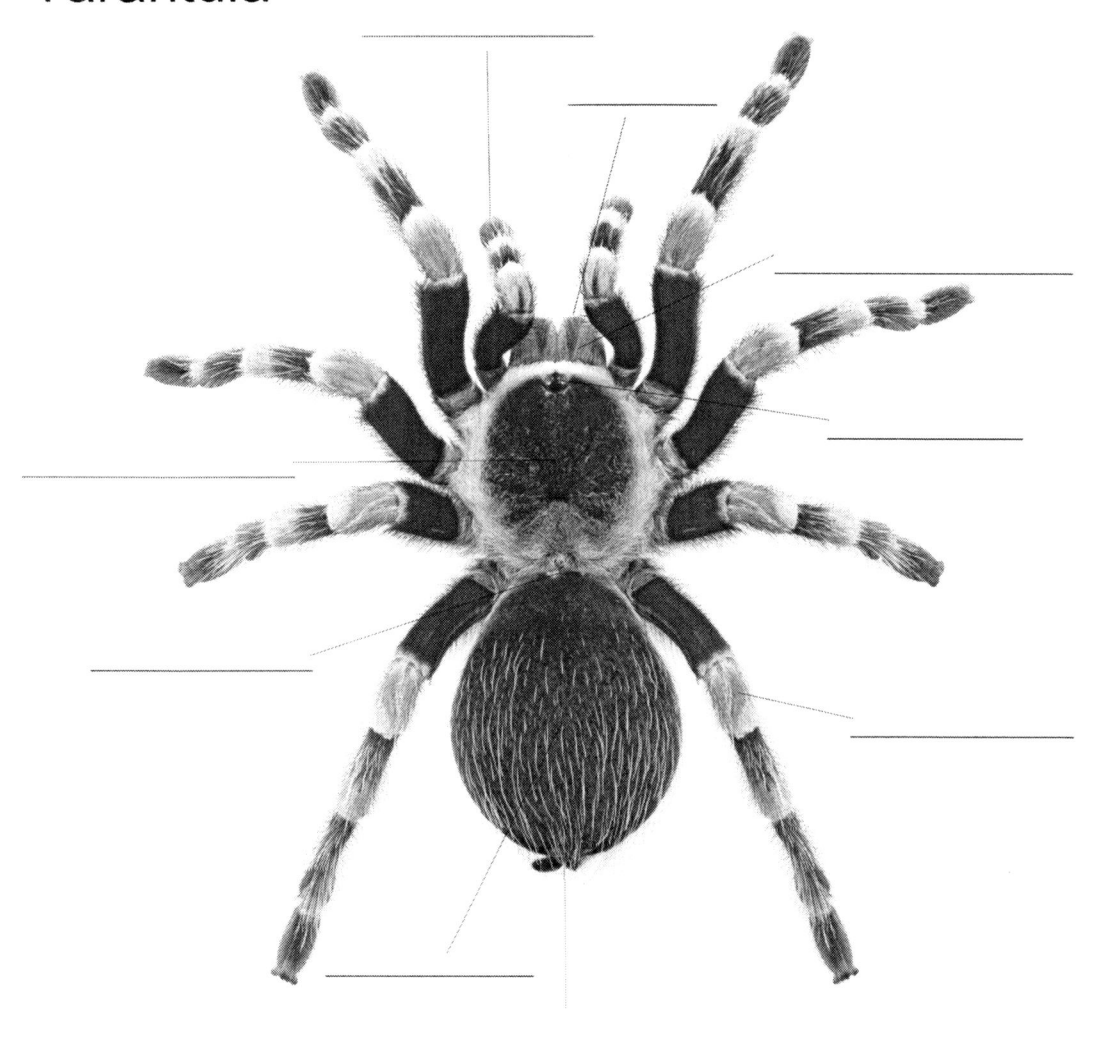

Label the horizontal lines above to denote these organs:

Eyes (8)   Spinnerets (4)   Legs (8)   Pedipalps (2)
Abdomen   Chelicerae (2)   Fangs (2)   Pedicel
Cephalothorax

© 2010 Classical Education Resources, LLC.   All rights reserved.

LAB 22: FOCUS ON A FROG

Diagram 22-1

External Anatomy of the Frog

Label the horizontal lines above to denote these features:

Hind limb
Pupil
Fore limb

Nictitating membrane
External nares
Tympanic membrane

LAB DIAGRAMS

| LAB 22: FOCUS ON A FROG | Diagram 22-2 |

Internal Anatomy of the Frog

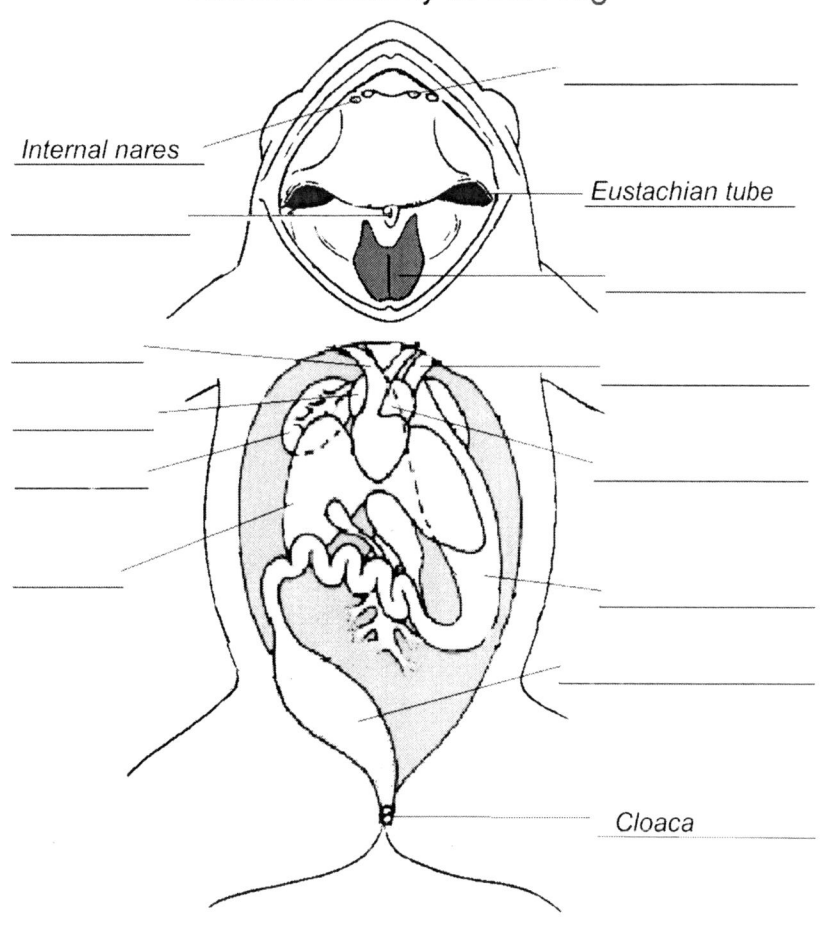

Internal nares

Eustachian tube

Cloaca

Label the horizontal lines above to denote these organs:

Glottis	Stomach	Lung	Vein
Tongue	Small intestine	Right atrium	Large intestine
Liver	Left atrium	Artery	Vomerine teeth
Ventricle			

© 2010 Classical Education Resources, LLC.   All rights reserved.

LAB DIAGRAMS

LAB 22: FOCUS ON A FROG — Diagram 22-3

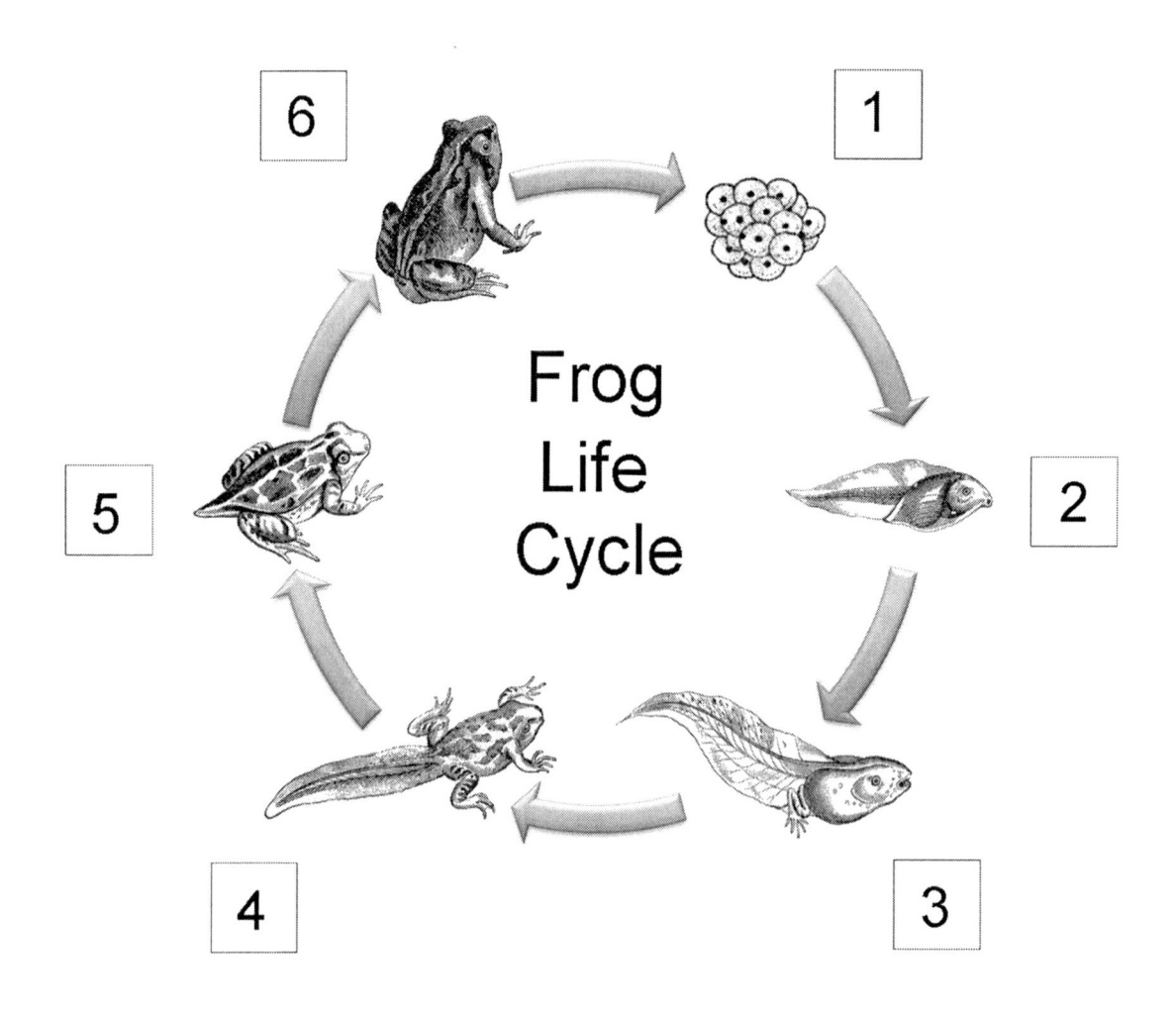

Frog Life Cycle

1. (spawn)
2.
3.
4.
5.
6.

Label the metamorphic stages below with the appropriate number:

_____ Tadpole with forelegs   _____ Adult frog

_____ Spawn (fertilized eggs)   _____ Froglet with diminishing tail

_____ Tadpole with hind legs   _____ Tadpole with external gills

© 2010 Classical Education Resources, LLC.   All rights reserved.

LAB DIAGRAMS

| LAB 23: CONSIDER A COBRA | Diagram 23-1 |

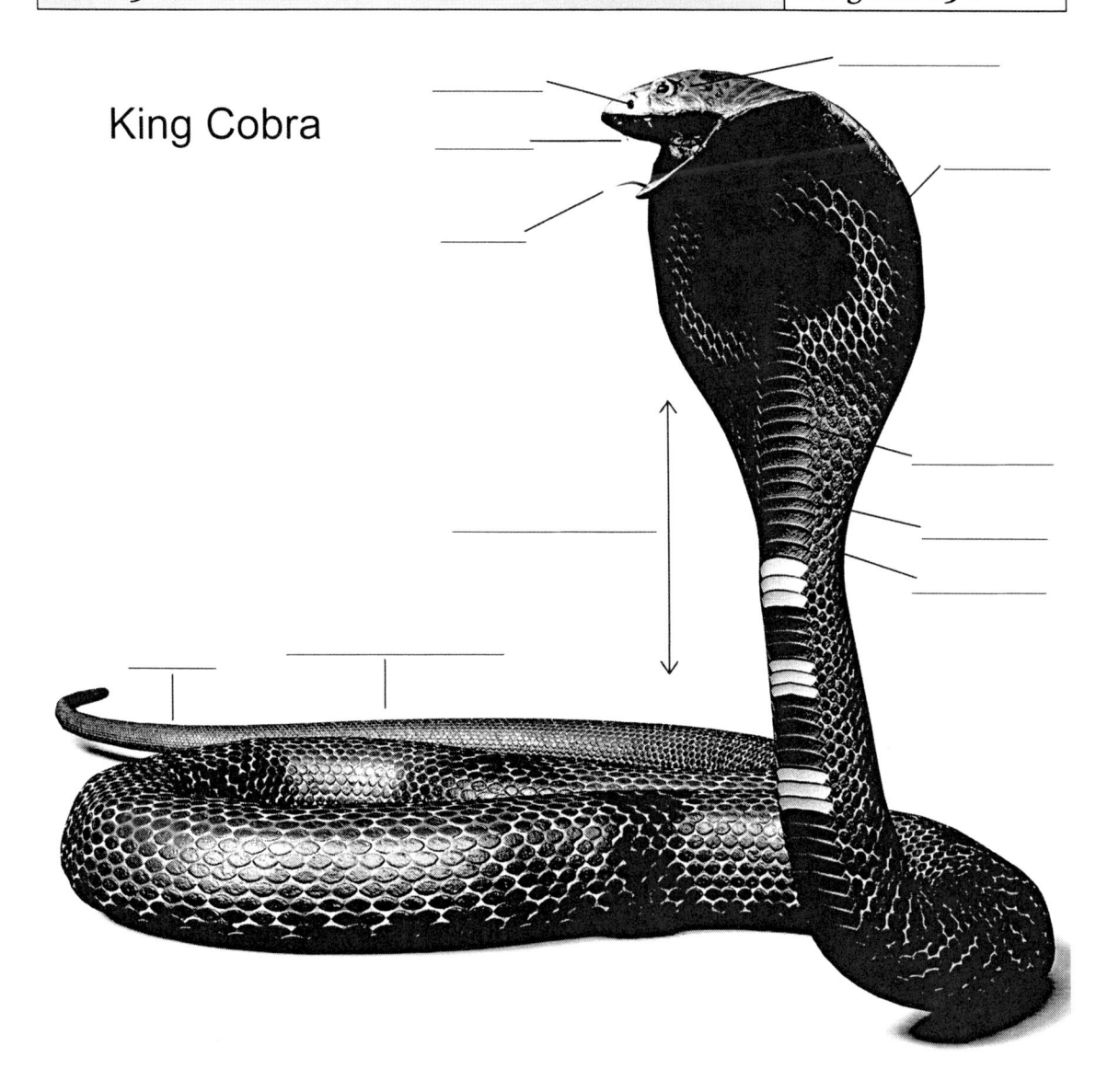

King Cobra

Label the horizontal lines above to denote the approximate locations of these organs:

Hood	Left lung	Vent	Nostril
Tracheal lung	Uterus & eggs	Fangs	Right lung
Heart	Venom gland	Tongue	

Remember: The right lung is very long.

© 2010 Classical Education Resources, LLC.   All rights reserved.

LIST A: THE SEVEN MAJOR GROUPS OF CLASSICAL TAXONOMY

- Kingdom

- Phylum/Division

- Class

- Order

- Family

- Genus

- Species

LIST B: THE FIVE KINGDOMS OF CLASSICAL TAXONOMY

- Protista
- Monera
- Fungi
- Plantae
- Animalia

LIST C: PARTS OF A FLOWER

- Receptacle
- Petals
- Sepals
- ~~Nectarines~~ Nectaries
- Stamens
- Pistil

LIST D: ANIMAL KINGDOM PHYLA

- Chordata

- Echinodermata

- Arthropoda

- Mollusca

- Annelida

- Nematoda

- Platyhelminthes

- Coelenterata

- Porifera

LIST E: MAJOR SYSTEMS OF THE HUMAN BODY

- Integumentary
- Skeletal
- Muscular
- Digestive
- Respiratory
- Circulatory
- Urinary
- Reproductive
- Endocrine
- Nervous
- Lymphatic
- Immune

LIST F: MAJOR BONES OF THE HUMAN BODY

- Cranium
- Mandible
- Clavicle
- Scapula
- Rib cage
- Ulna
- Radius
- Pelvis
- Carpus
- Femur
- Patella
- Fibula
- Tibia

LIST G: MAJOR BIOMES OF THE EARTH

- Tundra
- Coniferous forest
- Deciduous forest
- Tropical forest
- Temperate grassland
- Savanna
- Desert

RESOURCES

LAB REPORT: OBJECTIVE & MATERIALS

| LAB: | Date: |

OBJECTIVE

_____
_____
_____
_____
_____
_____
_____

MATERIALS

_____	_____
_____	_____
_____	_____
_____	_____
_____	_____
_____	_____
_____	_____
_____	_____
_____	_____
_____	_____

© 2010 Classical Education Resources, LLC.   All rights reserved.

LAB REPORT: PROCEDURE

| LAB: | Date: |

PROCEDURE

| LAB: | Date: |

HYPOTHESIS

LAB REPORT: RESULTS, NARRATIVE

LAB: | **Date:**

RESULTS

LAB REPORT: RESULTS, DATA

| LAB: | Date: |

RESULTS

NOTES

© 2010 Classical Education Resources, LLC.   All rights reserved.

LAB REPORT: RESULTS, MICROSCOPY

LAB:　　　　　　　　　　　　　　　　　**Date:**

RESULTS

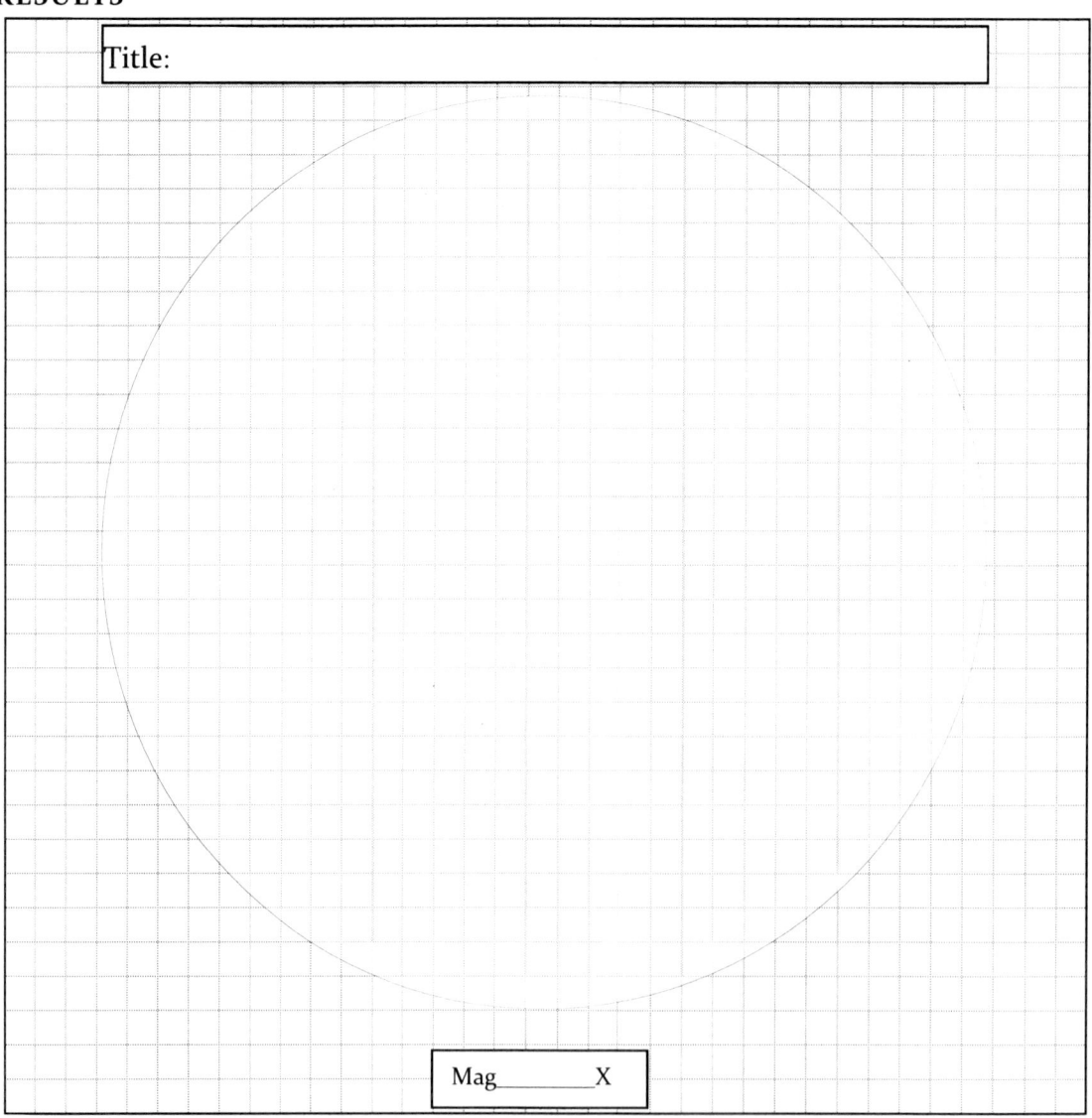

Title:

Mag_____X

NOTES

© 2010 Classical Education Resources, LLC.　All rights reserved.　　　~ 198 ~

LAB REPORT: RESULTS, OBSERVATION

LAB: | **Date:**

RESULTS

Title:

NOTES

LAB REPORT: CONCLUSION

LAB: | **Date:**

CONCLUSION

IMPORTANT DATES

ANCIENT (5000 BCE – 400 CE)

IMPORTANT DATES

MEDIEVAL – EARLY RENAISSANCE (400 – 1600)

IMPORTANT DATES

LATE RENAISSANCE-EARLY MODERN (1600 – 1850)

IMPORTANT DATES

MODERN (1850 – PRESENT)

TIMELINE

Prefix	Meaning	Origin
ante-	before	Latin
anti-	against	Greek
bi-	two	Greek
co-	together	Latin
com-	together	Latin
contra-	against, facing	Latin
di-	two	Greek
dia-	through, across, over	Greek
endo-	within, internal	Greek
epi-	upon, over, at, near	Greek
ex-	out, off, away, thoroughly	Latin
exo-	outside, external	Greek
extra-	outside	Latin
hyper-	over	Greek
hypo-	under	Greek
in-	not	Latin
inter-	between	Latin
intra-	within	Latin
macro-	large	Greek
magni-	great, large	Latin
meta-	among, between	Greek
micro-	small	Greek
mon-	one	Greek
multi-	many	Latin
pan-	all	Greek
para-	beside, beyond	Greek
peri-	around	Greek
poly-	many	Greek
post-	after	Latin
pre-	before	Latin

pro-	in favor of, forward, instead of	Latin
pseud-	false	Greek
re-	again, back	Latin
semi-	half	Latin
sub-	beneath, secretly	Latin
super-	above	Latin
sym-	with, together	Greek
syn-	with, together	Greek
tele-	at a distance	Greek
trans-	across, over	Latin
in-	in	Latin
tri-	three	Greek

Stem/Root/Word part	English	Origin
amphi	on both sides	Greek
arachn	spider	Greek
arthr(o)	joint	Greek
artic	joint	Latin
atri	main room	Latin
audi	hearing	Latin
bac	rod-shaped	Latin
bio	life	Greek
capill	hair	Latin
chloro	green	Greek
chrom	color	Greek
cochl	shell	Greek
cotyl	cup	Greek
crusta	shell, crust	Latin
cut(i)	skin	Latin
cycl	wheel	Greek
cyt(o)	cell	Greek

derm	skin	Greek
echino	spiny	Latin
eco	house	Greek
erythr(o)	red	Greek
fer	carry	Latin
gam	marriage	Greek
gen	race, kind	Latin
germin	sprout	Latin
gest	carry	Latin
homeo	like	Greek
homio	like	Greek
homo	like	Greek
horm	that which excites	Greek
incus	anvil	Latin
integ	covering	Latin
leuc(o)	white	Greek
leuk(o)	white	Greek
lingu	tongue	Latin
logy	study of	Greek
malleus	hammer	Latin
mamm	breast	Latin
mei	less	Greek
mit	thread	Greek
moll	soft	Latin
morph	form, shape	Greek
nas	nose	Latin
nom	name	Latin
nuc	nut	Latin
ocul	eye	Latin
olfact	smell	Latin
onym	a name	Greek

or	mouth	Latin
organ	tool, instrument	Greek
ot	ear	Greek
ov	egg	Latin
pend	hang	Greek
photo	light	Greek
phyll	leaf	Greek
plas	form, mold	Greek
pod	foot	Greek
poikilo	various	Greek
prot	first	Greek
rep	crawl, creep	Latin
rept	crawl, creep	Latin
ret	net	Latin
sci	know	Latin
scope	see, look at	Greek
sect	cut	Latin
seg	cut	Latin
sito	food	Greek
soma	body	Greek
spec	look	Latin
spher	ball	Greek
spir	breathe	Latin
tax	arrangement, order	Greek
therm	heat	Greek
thes	to put or place	Greek
thorax	chest, breastplate	Greek
trop	turning	Greek
ven	vein	Latin
vent	wind	Latin
ventr	belly	Latin

volunt	will	Latin
zyg	yoke	Greek

Suffix	Meaning	Origin
-ast	one associated with	Greek
-ia	pertaining to, pathological condition	Greek
-ics	thing having to do with	Greek
-ion	state or process	Latin
-ism	action, condition	Greek
-ist	advocates, makes	Greek
-ite	adherent, body part, rock/mineral	Greek
-ium	the act	Latin
-ment	result or means of an act	Latin
-oid	resembling	Greek
-or	one who performs action	Latin

ANSWER KEY

LAB 02: CATEGORICALLY SPEAKING   Date:

RESULTS

1. What different characteristics do you notice about the specimens? *shape, size, color*

2. How many specimens are large? *twelve*

3. How many specimens are blue? *six*

4. How many specimens are triangles? *eight*

5. How many specimens are both large and blue? *three*

6. How many specimens are large blue triangles? *one*

CLASSIFICATION POSSIBILITIES: *answers will vary, but there should be six different schemes – one possible example is shown*

First by...	Then by...	Finally by...
shape	color	size
shape	size	color
color	shape	size
color	size	shape
size	shape	color
size	color	shape

© 2010 Classical Education Resources, LLC.   All rights reserved.

ANSWER KEY

LAB 02: CATEGORICALLY SPEAKING

Date:

RESULTS

7. Does one classification system seem more logical than the others? Why or why not? *answers will vary; student only needs to recognize that more than one system is possible; if they prefer one system, they should give a logical reason why*

8. Look carefully at all the specimen cards. Can you discover other features that could be used to classify the organisms? *some possibilities to consider: the squares and triangles have angles/corners; the squares and circles have complete symmetry from left to right and top to bottom, whereas the triangles are only symmetrical from left to right; some shapes have "warm" colors, some have "cool" colors, etc.*

ANSWER KEY

LAB 02: CATEGORICALLY SPEAKING   Date:

RESULTS   *The layout shown below is one possibility; note that in this example there is only one item per group in the last column because we do not have any duplicates.*

In the two middle columns, draw lines to divide your specimens into groups.

© 2010 Classical Education Resources, LLC.   All rights reserved.   ~ 215 ~

ANSWER KEY

LAB 04: READY FOR MY CLOSEUP — Diagram 04-1

Compound Microscope

- Eyepiece (Ocular)
- Eyepiece tube
- Nosepiece
- Objectives
- Stage clips
- Stage
- Diaphragm
- Illumination Source
- Arm
- Focus knob

© 2010 Classical Education Resources, LLC.   All rights reserved.

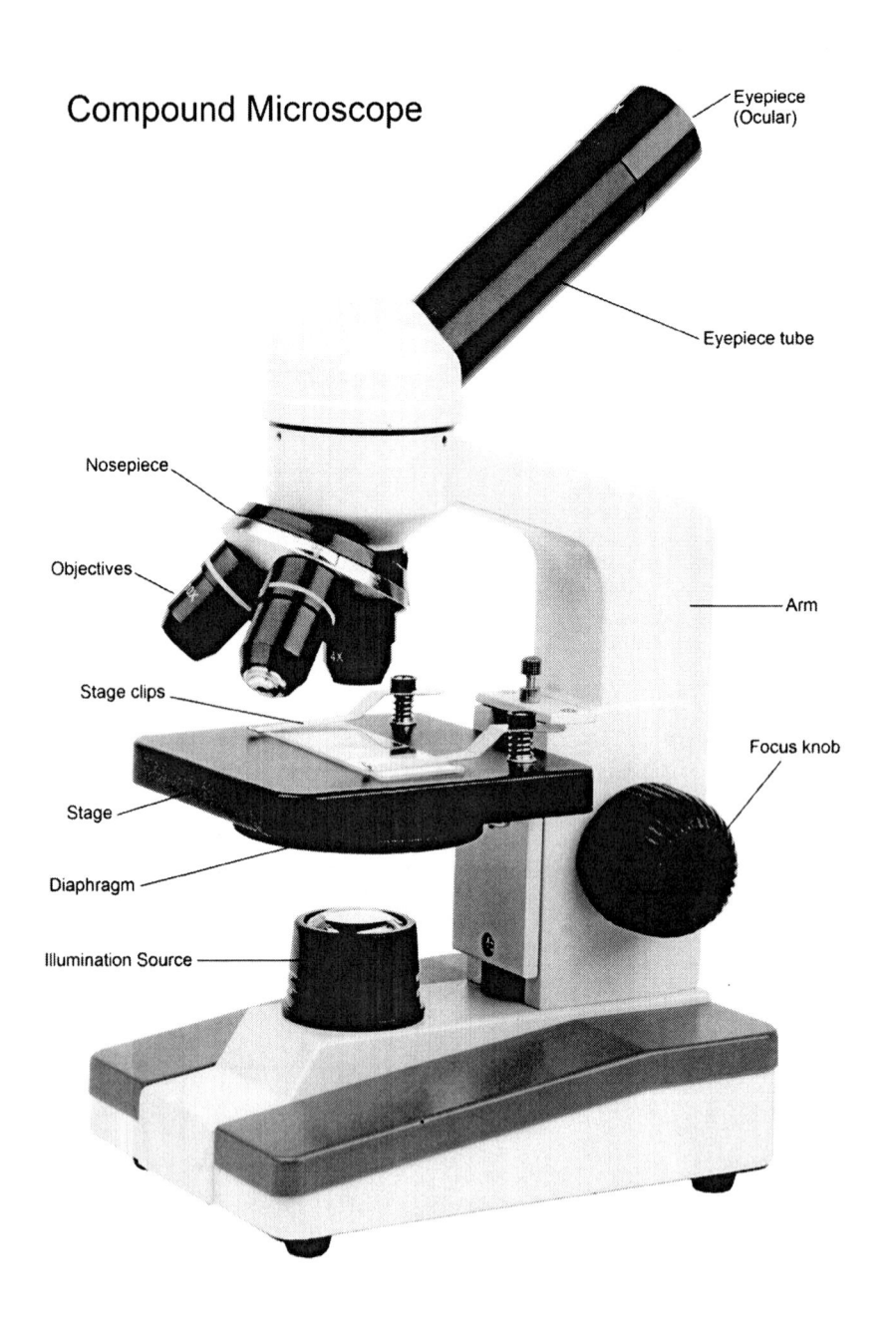

ANSWER KEY

| LAB 05: SAME DIFFERENCE | Handout 05-1 |

Plant Cell

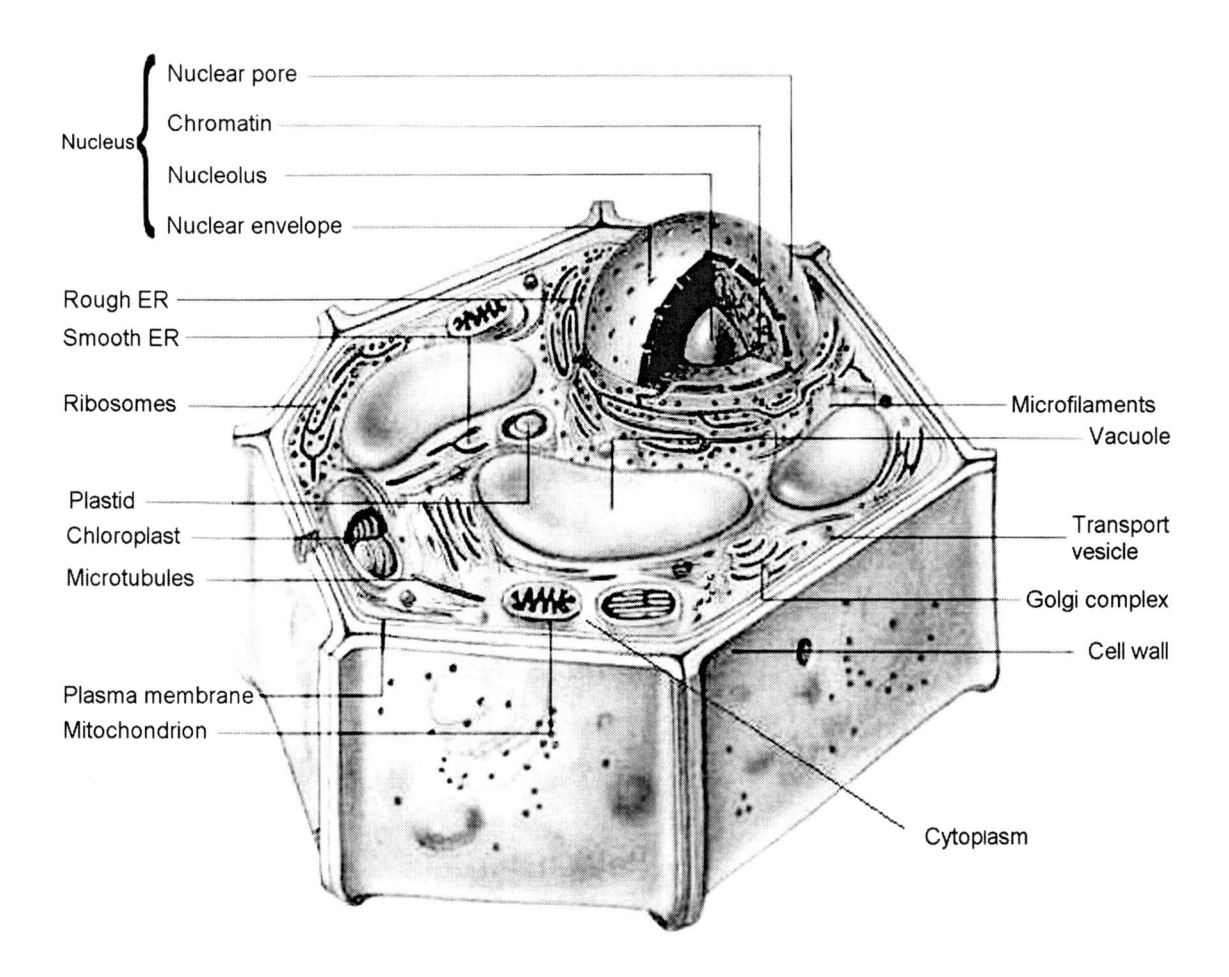

Nucleus
- Nuclear pore
- Chromatin
- Nucleolus
- Nuclear envelope

Rough ER
Smooth ER

Ribosomes

Plastid
Chloroplast
Microtubules

Plasma membrane
Mitochondrion

Microfilaments
Vacuole

Transport vesicle
Golgi complex
Cell wall

Cytoplasm

Label the horizontal lines above to denote these organelles:
- A Cell wall
- B Smooth ER
- C Cell membrane
- D Nucleus
- E Vacuole
- F Cytoplasm
- G Ribosome
- H Chloroplast
- I Rough ER
- J Mitochondrion
- K Golgi complex

*ER = Endoplasmic reticulum   Cell membrane is also known as plasma membrane

© 2010 Classical Education Resources, LLC.   All rights reserved.   ~ 217 ~

ANSWER KEY

LAB 05: SAME DIFFERENCE   Handout 05-2

Animal Cell

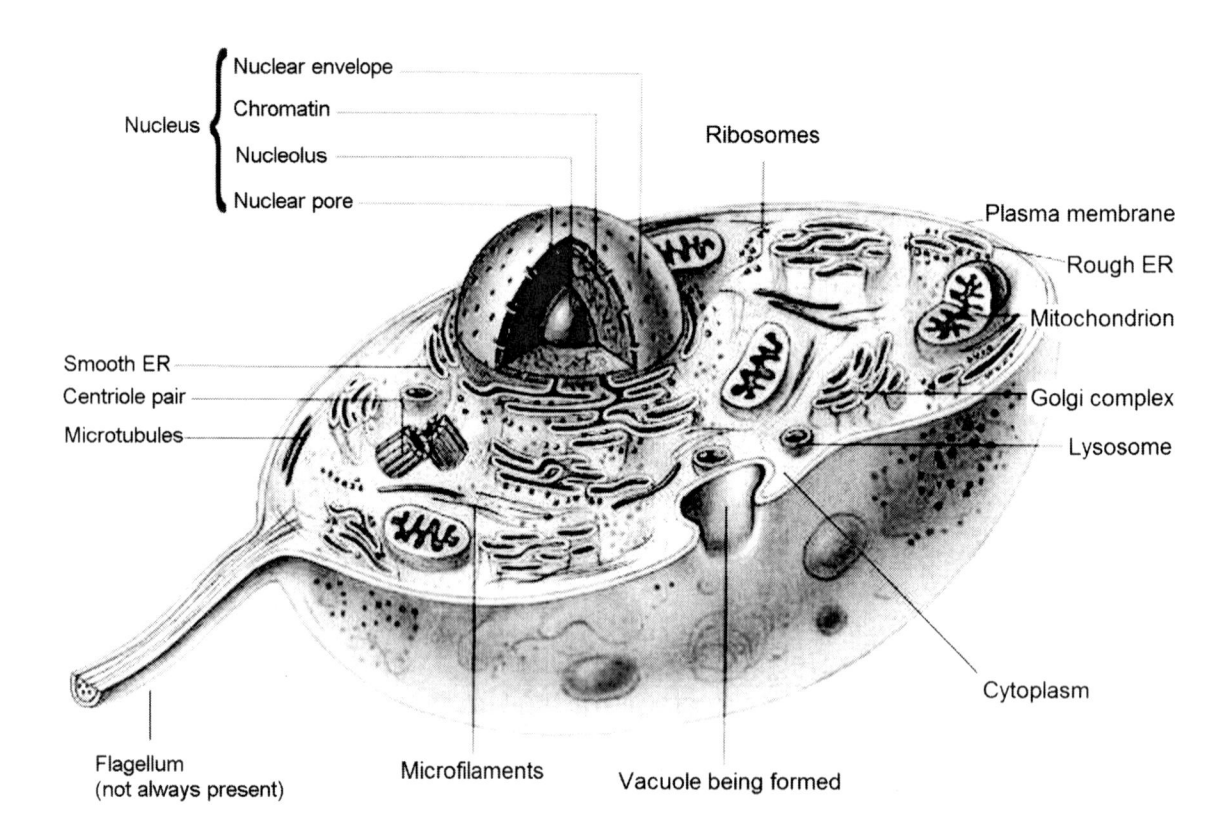

Labels on diagram:
- Nucleus: Nuclear envelope, Chromatin, Nucleolus, Nuclear pore
- Ribosomes
- Plasma membrane
- Rough ER
- Mitochondrion
- Golgi complex
- Lysosome
- Cytoplasm
- Vacuole being formed
- Microfilaments
- Flagellum (not always present)
- Microtubules
- Centriole pair
- Smooth ER

Label the horizontal lines above to denote these organelles:

A	Lysosome	E	Mitochondrion	H	Rough ER
B	Cell membrane	F	Ribosome	I	Nucleus
C	Smooth ER	G	Vacuole	J	Cytoplasm
D	Golgi complex				

* ER = Endoplasmic reticulum   Cell membrane is also known as plasma membrane
* In this picture, vacuole is being formed on edge of cell

© 2010 Classical Education Resources, LLC.   All rights reserved.   ~ 218 ~

ANSWER KEY

LAB 07: SPLIT DECISION Diagram 07-1

Mitosis

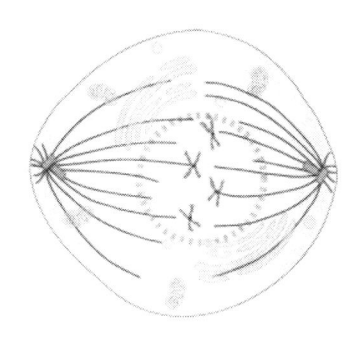

Prophase

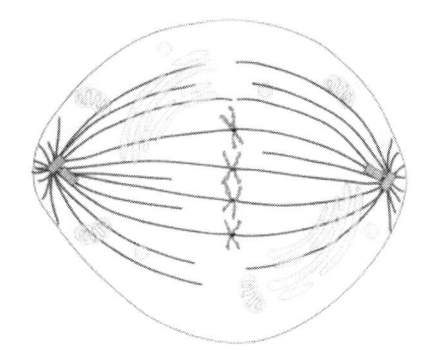

Metaphase

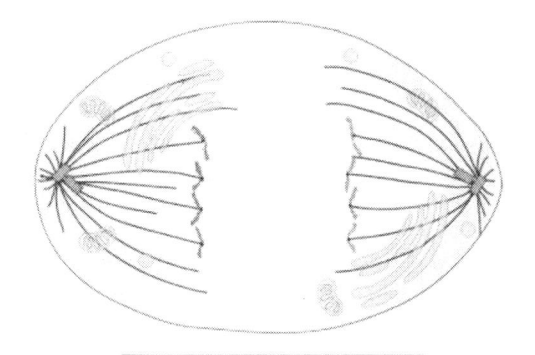

Anaphase

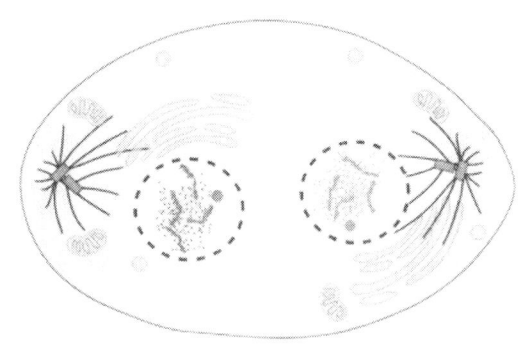

Telophase

© 2010 Classical Education Resources, LLC.   All rights reserved.

ANSWER KEY

LAB 07: SPLIT DECISION — Diagram 07-2

Meiosis I

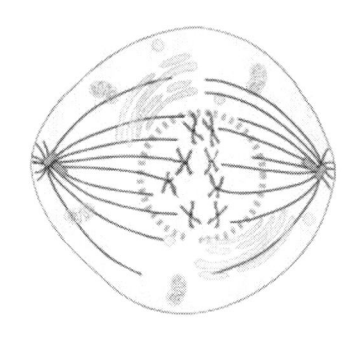

Prophase I

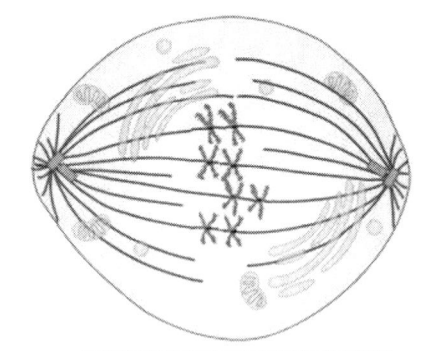

Metaphase I

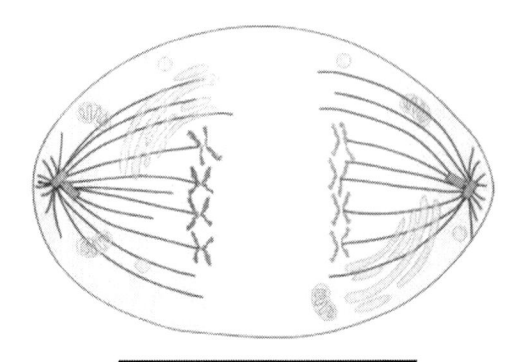

Anaphase I

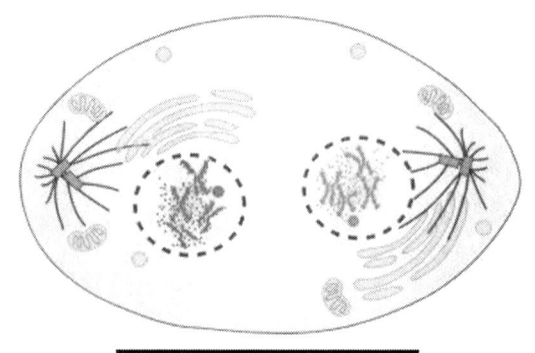

Telophase I

© 2010 Classical Education Resources, LLC.   All rights reserved.

ANSWER KEY

LAB 07: SPLIT DECISION — Diagram 07-3

Meiosis II

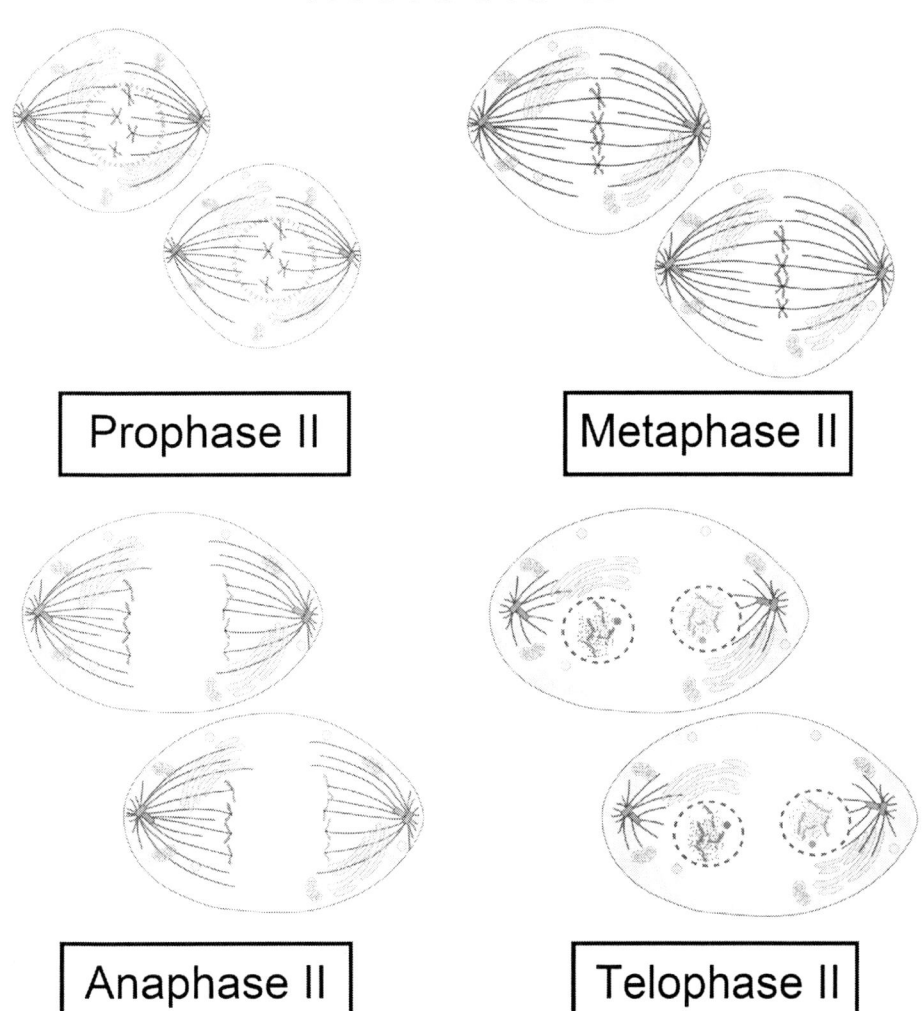

Prophase II

Metaphase II

Anaphase II

Telophase II

© 2010 Classical Education Resources, LLC.   All rights reserved.

ANSWER KEY

LAB 08: A NEW PAIR OF GENES   Date:

RESULTS

1. What type of hair do the example parents above have? Remember their genotypes are both Cc. _curly_

2. What if one parent has both genes Cc (hybrid) and the other has two dominant genes CC? Fill in the Punnett Square and determine the probability of curly hair in the children.

Mother = Cc

	C	c
C	CC	Cc
C	CC	Cc

Father = CC

% chance of curly hair (C is present in square) _____ _100%_

% chance of straight hair (no C is present in square) _____ _0%_

What are the possible genotypes for these children? _____ _CC or Cc_

3. What parental genotypes would guarantee only straight-haired children?

4. Father ___cc___   Mother ___cc___

5. Now let's have some fun. Edward and Bella have 400 (!) children. Using the following traits and genotypes, figure approximately how many children would express each trait.

Trait	Dominant	Recessive	Edward	Bella
Eye color	B=brown	b=blue	BB	bb
Hair color	D=dark	d=light	dd	Dd
Hair texture	C=curly	c=straight	Cc	Cc

© 2010 Classical Education Resources, LLC.   All rights reserved.

ANSWER KEY

| LAB 08: A NEW PAIR OF GENES | Date: |

RESULTS

Eye color:

Edward = BB
Bella = bb

	b	b
B	Bb	Bb
B	Bb	Bb

Hair color:

Edward = dd
Bella = Dd

	D	d
d	Dd	dd
d	Dd	dd

Hair type:

Edward = Cc
Bella = Cc

	C	c
C	CC	Cc
c	Cc	cc

6. Approximately how many children will have brown eyes? __400 (100%)__
7. Approximately how many children will have blue eyes? __0 (0%)__
8. Approximately how many children will have dark hair? __200 (50%)__
9. Approximately how many children will have light hair? __200 (50%)__
10. Approximately how many children will have curly hair? __300 (75%)__
11. Approximately how many children will have straight hair? __100 (25%)__
12. Their last (!) child has the following genotypes:

    Bb   dd   CC

13. What does she look like? __brown eyes and light, curly hair__

    For eyes, brown is dominant and blue is recessive. For hair color, dark is dominate and blond is recessive.

14. Can you guess your genotypes for eye and hair color? __answers will vary__

15. Can you guess the possibilities for your parents' genotypes?

    Father's eye color __answers will vary__   Father's hair color __answers will vary__

    Mother's eye color __answers will vary__   Mother's hair color __answers will vary__

© 2010 Classical Education Resources, LLC.   All rights reserved.

ANSWER KEY

LAB 19: STUDY A SPIDER — Diagram 19-1

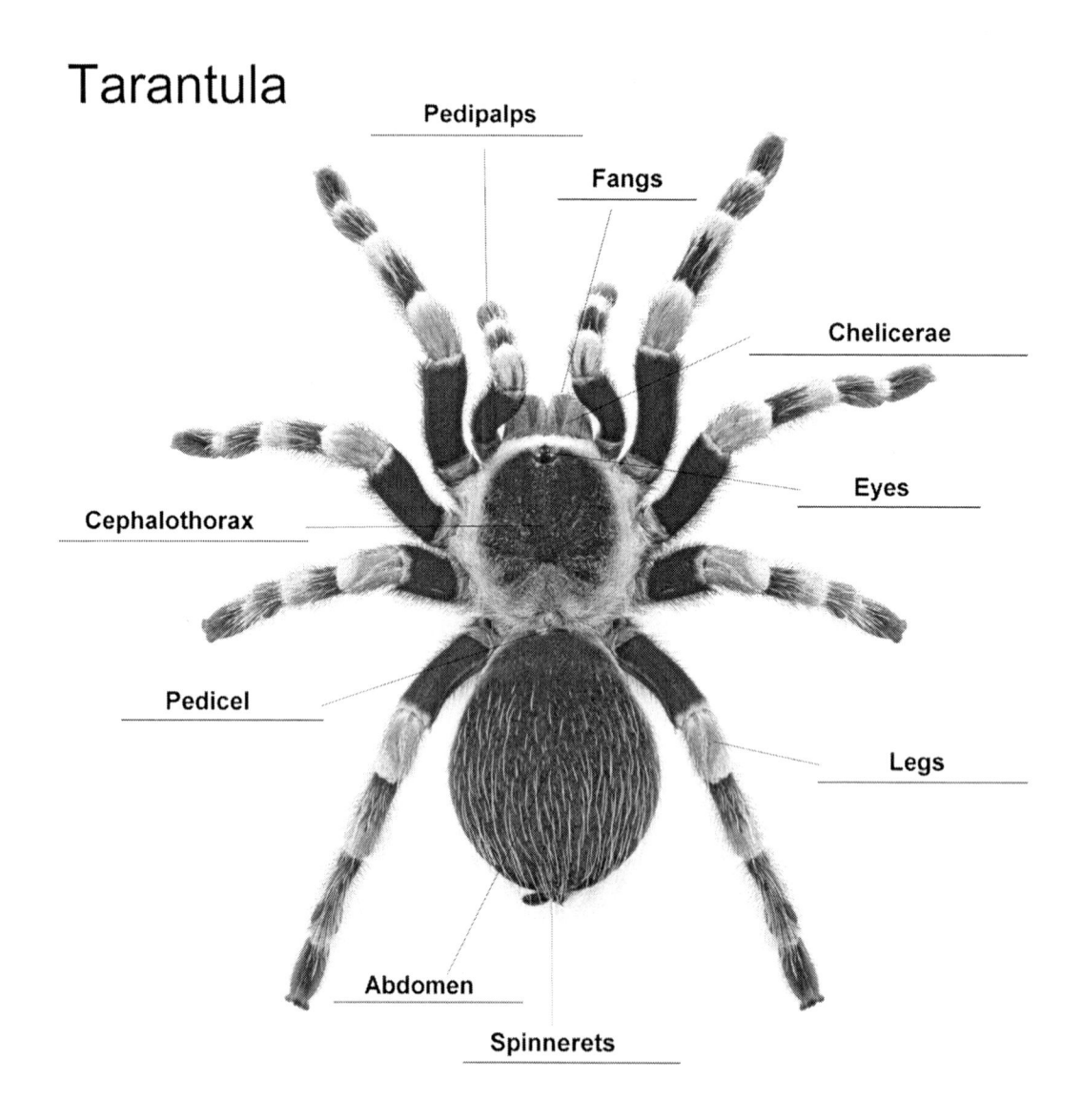

Tarantula

- Pedipalps
- Fangs
- Chelicerae
- Eyes
- Cephalothorax
- Pedicel
- Legs
- Abdomen
- Spinnerets

Label the horizontal lines above to denote these organs:

Eyes (8)	Spinnerets (4)	Legs (8)	Pedipalps (2)
Abdomen	Chelicerae (2)	Fangs (2)	Pedicel
Cephalothorax			

© 2010 Classical Education Resources, LLC.   All rights reserved.

~ 224 ~

ANSWER KEY

LAB 22: FOCUS ON A FROG — Diagram 22-1

External Anatomy of the Frog

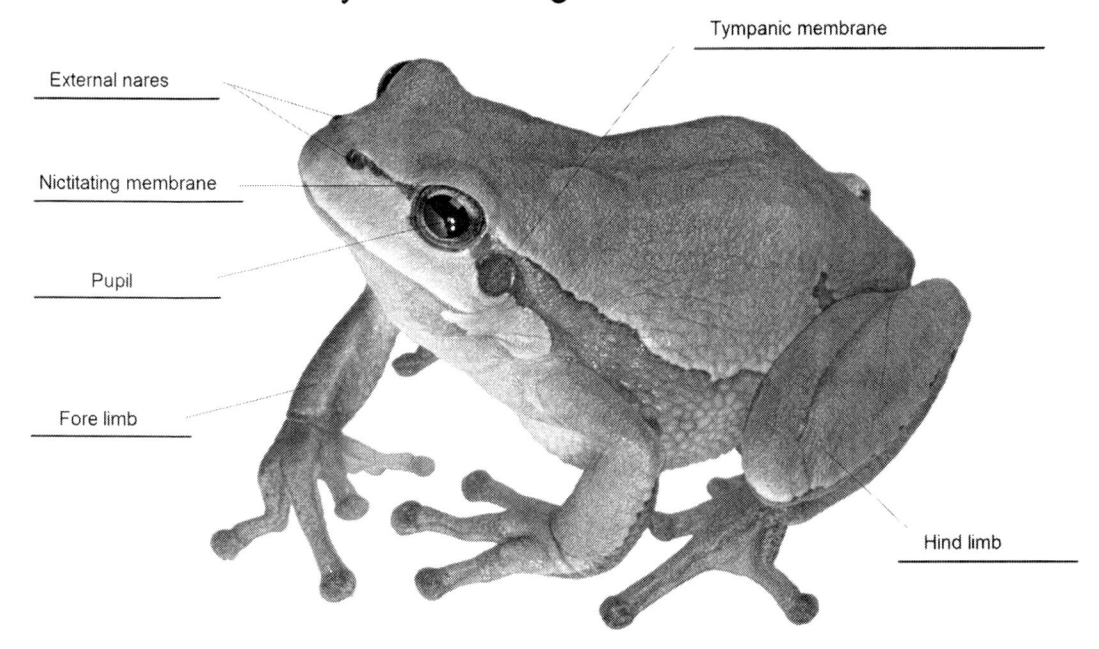

- Tympanic membrane
- External nares
- Nictitating membrane
- Pupil
- Fore limb
- Hind limb

Label the horizontal lines above to denote these features:

Hind limb   Nictitating membrane
Pupil   External nares
Fore limb   Tympanic membrane

© 2010 Classical Education Resources, LLC.   All rights reserved.

ANSWER KEY

LAB 22: FOCUS ON A FROG — Diagram 22-2

Internal Anatomy of the Frog

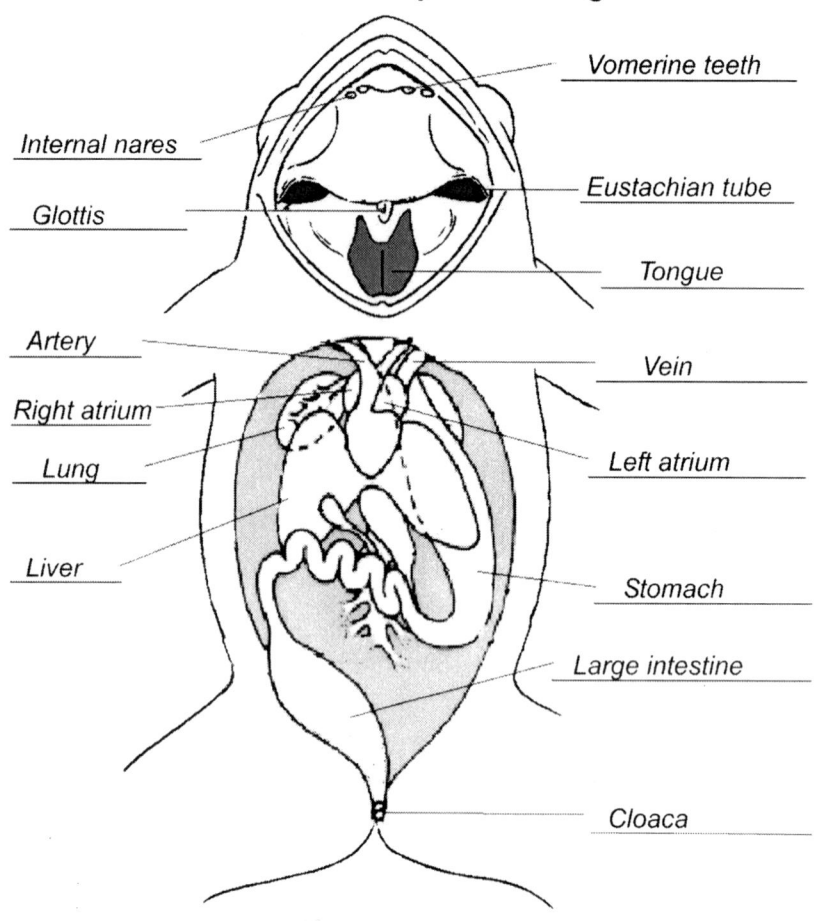

Labels on diagram:
- Vomerine teeth
- Internal nares
- Eustachian tube
- Glottis
- Tongue
- Artery
- Vein
- Right atrium
- Left atrium
- Lung
- Liver
- Stomach
- Large intestine
- Cloaca

Label the horizontal lines above to denote these organs:

Glottis	Stomach	Lung	Vein
Tongue	Small intestine	Right atrium	Large intestine
Liver	Left atrium	Artery	Vomerine teeth
Ventricle			

© 2010 Classical Education Resources, LLC.   All rights reserved.

ANSWER KEY

LAB 22: FOCUS ON A FROG — Diagram 22-3

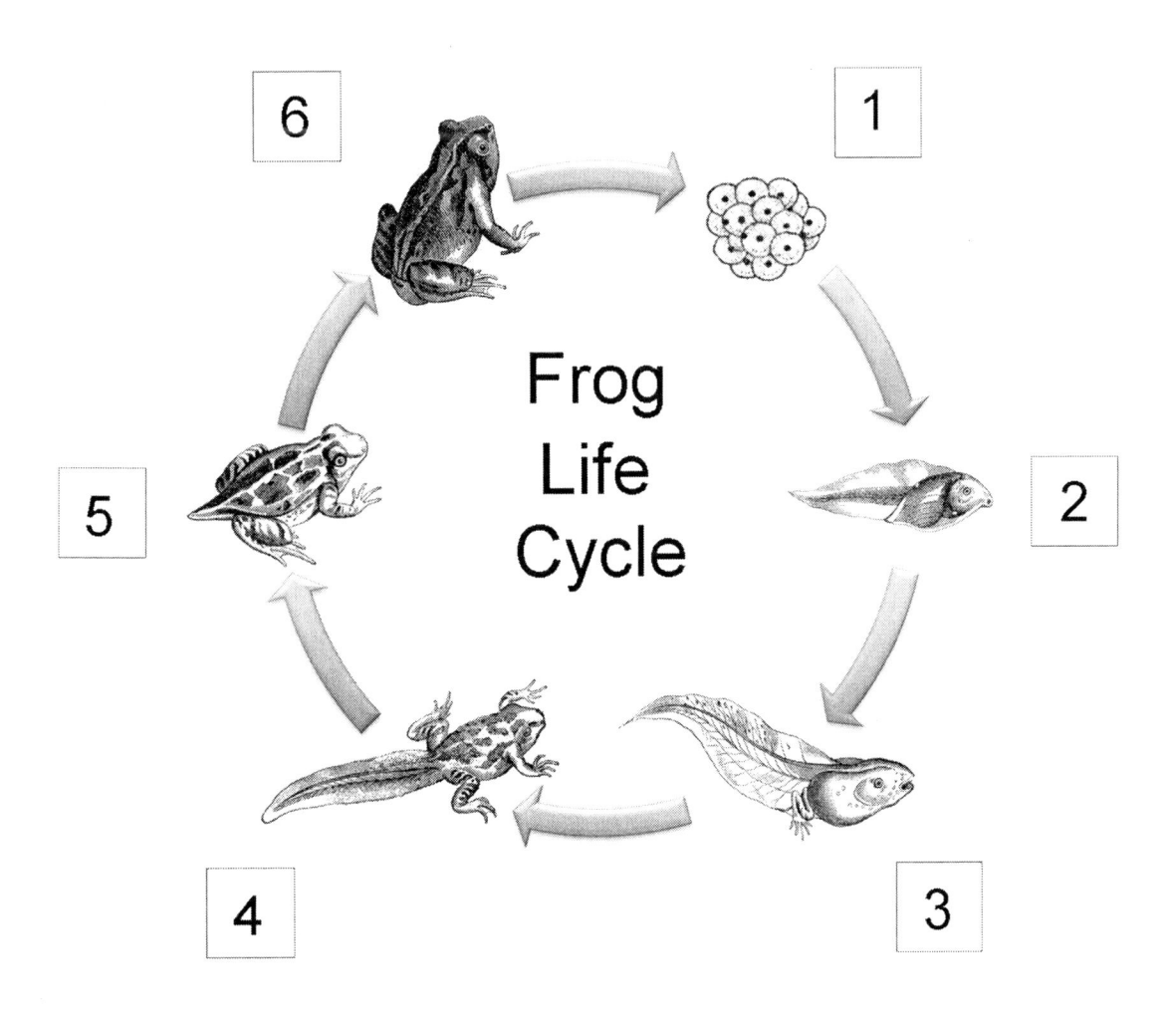

Frog Life Cycle

Label the metamorphic stages below with the appropriate number:

__4__ Tadpole with forelegs

__1__ Spawn (fertilized eggs)

__3__ Tadpole with hind legs

__6__ Adult frog

__5__ Froglet with diminishing tail

__2__ Tadpole with external gills

© 2010 Classical Education Resources, LLC.   All rights reserved.

ANSWER KEY

LAB 23: CONSIDER A COBRA — Diagram 23-1

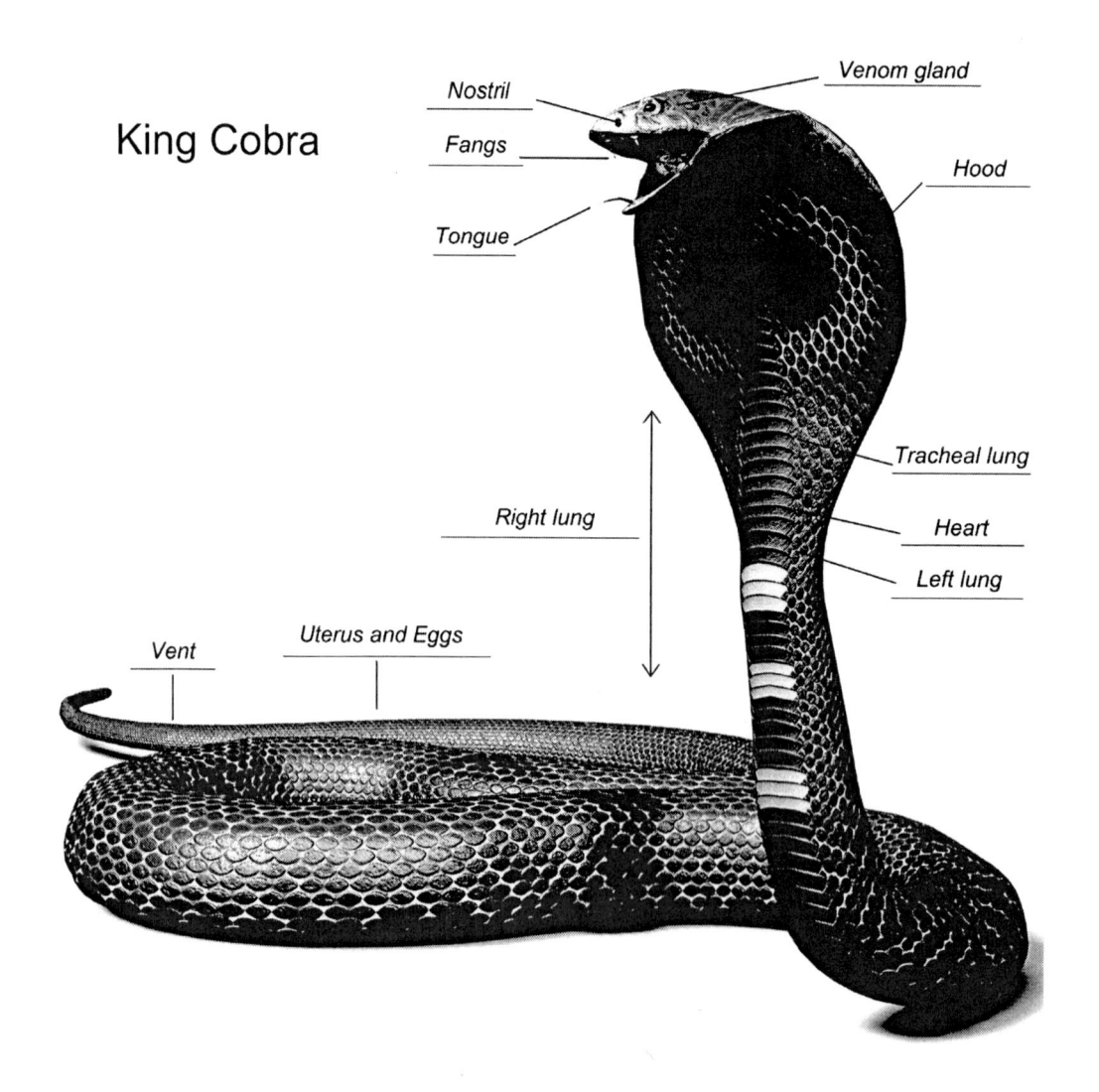

King Cobra

- Nostril
- Fangs
- Tongue
- Venom gland
- Hood
- Tracheal lung
- Heart
- Left lung
- Right lung
- Vent
- Uterus and Eggs

Label the horizontal lines above to denote the approximate locations of these organs:

Hood	Left lung	Vent	Nostril
Tracheal lung	Uterus & eggs	Fangs	Right lung
Heart	Venom gland	Tongue	

Remember: The right lung is very long.

© 2010 Classical Education Resources, LLC.   All rights reserved.

THINKING CRITICALLY ABOUT SCIENCE

The Power and Limits of Science

Can all questions be answered by using the methods of science? In the course of your studies you will encounter theories about scientific matters that are not really theories in the scientific sense. Ideas and beliefs are often put forth as scientific theories, but unless the idea is subject to scientific experimentation, it cannot be a *scientific* theory.

To illustrate this concept, think about someone that you love. How would you prove, in a scientific sense, that you love them? Perhaps you show your love by doing nice things for them. Does that prove that you actually love them? Not really, because you do nice things for lots of people, not just the ones you love. And history is full of examples of people who have behaved kindly toward another person while secretly plotting their downfall. So, observable actions are not adequate to prove that love is actually present.

But what if a researcher hooked electrodes up to your brain and measured your brain wave activity when you are near the person you love. The researcher might be able to detect some electrical changes and gather some data. Perhaps they could graph the data and come up with an impressive lab report. Would they have actually proven that you love the other person? In fact, all they have really shown is that there are some electrical changes in your brain when you are near the person. Those changes may be an effect of love, but love itself remains undetectable and unproven.

The reason for this is that love is intangible. It is not subject to *empirical* observation – it cannot be inspected with the physical senses. While it may be generally agreed that love exists, intangibles are not subject to scientific testing methods because they cannot be directly observed and measured. And for that reason, their existence can never be proven in the scientific sense.

You may encounter researchers who attempt to invalidate other people's beliefs by using scientific methods. But in order to be studied by science, a thing must be subject to physical experience. Intangibles, like beliefs and ideas, are not subject to scientific verification. They can neither be proven nor disproven.

Bigfoot and Space Aliens

Have you ever seen a live cobra? How about a Mongolian jird? Are they real?

There are people who believe that undiscovered primates are currently living in various parts of the earth. Scientists who study these types of proposed but unproven life forms are called cryptozoologists.[3] To date, there is not adequate evidence that creatures such as Bigfoot exist. Does that mean that the sasquatch is not real? What *can* we say about the existence of such animals? If this animal is ever captured and brought in to be studied we will be able to say that they are real. Until such time, can we say that they are not? Not really. All we can say is that there is not adequate evidence that they *do* exist.

Are extraterrestrial aliens visiting the earth? If a spaceship ever lands on the White House lawn we might be able to say that they are. In the mean time, since we cannot possibly be everywhere at once, we cannot say definitively that they are *not* visiting, only that we do not have adequate evidence that they *are* visiting.

Absence of evidence does not equal *evidence of absence*. Until we have direct experience of something, we can't say anything scientific about it. All we can say is that we do not know. So, just because something has not been proven to exist yet, does not necessarily mean that it does not exist. And just because something, like love, is not provable in a scientific sense, does not necessarily mean that it is not real.

Science is a profoundly useful tool for the study of the physical universe. But, it deals only with that which is empirical – directly observable through the senses. Things like ideas and beliefs are not usually suitable subjects for empirical scientific study.

If we are to aspire to wisdom we must walk a "razor's edge". It is important to remain open-minded and at the same time skeptical. To do this, we must keep in mind the difference between that which is knowable, that which is knowable but not yet known and that which can never be known in a scientific sense. Many mysteries are not yet solved, but may be solved one day. Many mysteries can only be addressed through faith.

"Don't believe everything you think."

Anonymous

---

[3] *crypto* means hidden, *zoology* is the study of animals; *cryptozoologists* study "hidden" animals

All images were either created by the author, exist in the public domain, or are used under royalty-free license from www.dreamstime.com

CPSIA information can be obtained at www.ICGtesting.com
Printed in the USA
LVOW09s1606210713

343907LV00001B/13/P

9 780982 957318